LEADER

Proven Habits for Sustainable Success in the Digital World

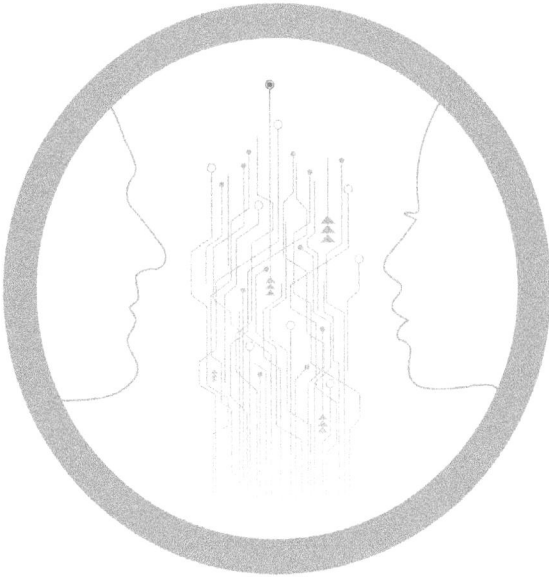

DEBASIS BHAUMIK

Editorial Project Management: Karen Rowe, www.karenrowe.com

Cover Design: Shake Creative, ShakeTampa.com

Inside Layout: Ljiljana Pavkov

Printed in the United States

ISBN: 978-1-7770689-0-5 (paperback)

ISBN: 978-1-7770689-1-2 (ebook)

Technology is a commodity;
people are the differentiators.

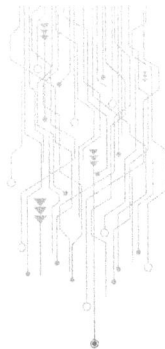

Table of

Contents

To my Dad,
an ordinary human with
extraordinary thoughts.

LEADERSHIP 4.0

Proven Habits for Sustainable Success in the Digital World

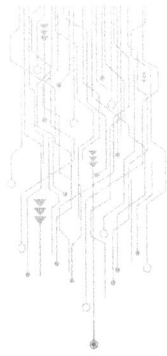

Introduction

We are in the middle of an incredibly interesting time in human evolution, particularly in our relationships with technology. In fact, we have made so much progress in technology that it is now challenging our roles in work and in society.

Some of us are very excited about these advances, as we believe that this challenging environment will make us stronger and better. We see limitless possibilities for growth in this time of massive change. However, the majority of our population is nervous about the challenges that these advances pose to their livelihood—and to some extent, about their role in general.

For millennia, humans have believed that our brains couldn't be replicated. We're astonished and fearful when we learn that, in some cases, technology does better, more efficient work than our brains do. The advent and increasing complexity of technology, especially artificial intelligence (AI), has left many humans feeling threatened. What's more, the media fuel our fears by insinuating that AI will soon become so advanced that it could *completely replace* our human workforce in mere years.

At the time of this writing, human beings are facing an unprecedented pandemic in the form of the COVID-19 crisis, which has disrupted our very way of life. Yet drones, self-driving vehicles and AI-enabled supply chain solutions are immensely helping mankind, working 24/7 without fatigue or fear. In the face of this crisis, technology is helping us in incredible ways, complementing our efforts in healthcare and social and individual initiatives.

Even though technology is immensely helping mankind, the sheer level of intelligence and near-human cognition of today's technology is mind-boggling to most people. The very concept of these spectacular advances compromises our sense of safety and security on multiple levels.

Organizations, like individuals, fear the potential of being replaced and overtaken by competition enabled by newer technology. Many corporations and organizations are still in denial about technology's role as the core of their business.

As a result, the leaders of these organizations are hesitant to invest in technology, yet they're also suffering from a fear of missing out (FOMO). They know they won't look good if they don't embrace technology on a larger scale, but they're not convinced that the results of implementing it can be effectively monetized. Generally speaking, leaders want to see a direct link to the bottom line.

Since traditional strategies and business forecasts are currently still working, corporate leaders seem to believe that traditional methods will always continue to work—whether or not their organizations become technology-centric. History has proven, however, that failing to innovate and reimagine business processes results in organizations being left behind.

In the midst of the COVID-19 crisis, the importance of adopting a technology-centric approach has become all the more obvious to me and to many others. It is amazing, for example, to see how thousands of schools and universities have shifted entirely to online classes. It has made me wonder whether there will still be a need for traditional universities at all twenty years from now. Even traditional organizations like those in heavy engineering, mining, manufacturing, and not-for-profit have shifted their operations to incorporate remote technology wherever possible.

On the other hand, lots of organizations are suffering. I have talked to many senior leaders who are cursing themselves for not further advancing their remote-operation agenda prior to this crisis. Some organizations are struggling just to enable their office workers to work from home, let alone sustain their industrial operations and supply chains.

This time, there might be an excuse—that this crisis is novel and unprecedented in our lifetime. But in the future, in the face of any crisis that disrupts business at the global or local level, there will be no excuses left. Organizations that are lagging in their readiness will surely be left behind.

The Fourth Industrial Revolution

I believe that the advancement of technology and human imagination has brought us to the most exciting juncture of human history thus far: an era that has come to be known as the **Fourth Industrial Revolution**, or **Industrial Revolution 4.0**.

Industrial Revolution 4.0 is wildly different from earlier industrial revolutions in human history. Previous industrial revolutions were characterized by automated jobs that required repetitive tasks and well-defined operating procedures. But we're changing. This industrial revolution is about **reimagining work and turning traditional procedures on their heads**. Now, we must reimagine what we're doing, how we're doing it, and the role of humans in this new ecosystem.

This process of reimagination of the workplace is exceedingly difficult. The path forward is anything but simple. Our options are to either jump in and commit to it with all our hearts or close our eyes in denial, for fear of being overwhelmed by change.

The biggest problem is that the status quo feels comfortable. Many leaders ask, "Why should I disrupt this? The money is coming, and the board is happy. We have a good forecast, and I'm happy now." In other words, they are content.

When there's no immediate stimulus to push us to make a change, our own comfort can sabotage our progress. Age and years of experience lead us to believe we're on the right path. We convince ourselves that the future will be bright, whether or not we embrace new technology. But the secret isn't in our years of experience, our age, or the time we've put into our industry. It's in our ability to be agile and to adapt along with the changes.

I started writing this book primarily with technology-driven disruption in mind, but the global pandemic has given us a wake-up call in another significant way. Sustainability in response to this pandemic will be one of the stimuli to move the technology agenda faster. Risks around operational sustainability highlighted by the COVID-19 crisis should provide a wake-up call for organizations. Reimagining our operations will be a key component of the larger overall agenda of reimagining business.

Reimagining our place in the market means getting out of our comfort zone. This book will provide some very simple recommendations for organizations and individuals to be successful during this disruptive time.

Leadership 4.0

I began my career in software development, and I have since worked with many companies and leadership teams. The irony is that, even though I started as a technologist, I believe that technology is a commodity. Technological advances like AI and quantum computing hold tremendous possibilities for the future; our ability to meet social needs, healthcare needs and other life essentials is getting better and better every day because of technology.

But everything comes with a trade-off.

Just like medicine, technology has side effects we must learn to respond to. But it also offers exponential rewards. In this book, I offer a silver bullet for any traditional organization seeking to transition to a technology-first organization. The solution is to **develop or upgrade your people to be digital leaders**.

In this book, I will use the words "leader" and "leadership" a lot. A leader, according to my definition, does not necessarily manage people. In the knowledge-based world, whether or not you manage people doesn't matter in the grand scheme of things. It's all about leading toward change or a specific outcome.

Whether you like it or not, every industry is becoming a knowledge-based, data-driven industry. While there are other underlying changes, the knowledge base is the primary source of this evolution.

Being a digital leader is much more about fundamentals than it is about a title. It's about how you show up at work—and outside of work. It's about how you live your life and how that affects your organization and your team. Your organization's status and your status as a digital leader are based on your knowledge and your behaviours.

In Part 2 of this book, I will talk about each of the above habits in detail and examine what it takes to be a digital leader on an individual level. But first, we'll start with gaining a brief understanding of the context. We'll take a look at the concept of Industrial Revolution 4.0, the "organizational lens," and, finally, how to become a digital leader.

Technology is vast and ever-changing. To survive in this new landscape, organizations and individuals must evolve at the same pace. The following chapters will show you how.

Instead of assuming and fearing our eventual demise at the hands of technology, we must imagine how we're going to coexist and thrive alongside it.

Part 1:

Digital Leadership
for Organizations

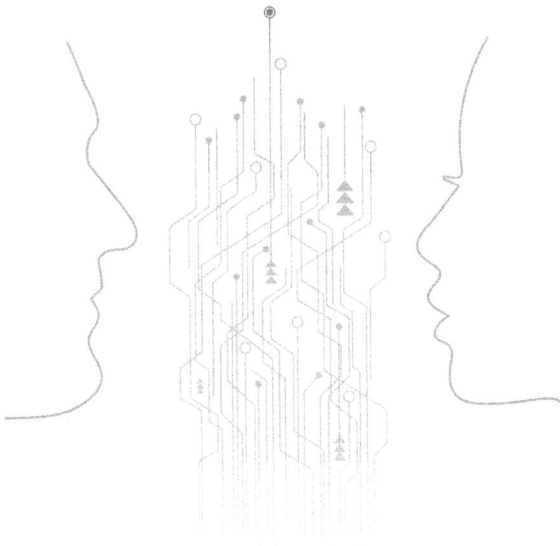

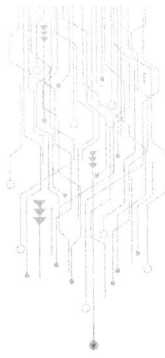

Chapter 1:
Context is Critical

"The Fourth Industrial Revolution, finally, will change not only what we do but also who we are. It will affect our identity and all the issues associated with it, from our sense of privacy, our notions of ownership, our consumption patterns, the time we devote to work and leisure, and how we develop our careers, cultivate our skills, meet people, and nurture relationships."[1]

—KLAUS SCHWAB

In 1943, a three-year-old girl approached her father with an innocent question. Her father, Edwin Land, was a photographer, and she wanted to know why she couldn't see the photographs he took immediately after they were taken. Why did she have to wait for them to be developed?

This innocent ask drove Dr. Land—the inventor of the Polaroid camera—to come up with a completely new process of photography, resulting in a camera that evolved and dominated household and professional photography for at least 30 years. Over that period of time, Polaroid sales grew from just $1.5 million to $1.5 billion.

The story of Edwin Land and his Polaroid[2] has always fascinated me—not just because the birth of household instant cameras started from a child's curious question, but because of what happened after that. Despite Polaroid's meteoric rise and decades-long dominance over the marketplace, it had a similarly dramatic downfall. It's a story that modern-day organizations would be wise to learn from, especially those seeking to position themselves as technology-first organizations.

Technology-First Organizations and Types of Innovation

Technology-first organizations (or TFOs) are organizations that consider technology to be at the core of all their business operations and transformation. These are the organizations that are choosing to reimagine work in the face of Industrial Revolution 4.0. The exciting thing about TFOs is that their technological evolution allows them to grow numerous offshoot businesses from their initial core offering.

TFOs innovate in two ways: product innovation (P-type innovation) and strategy innovation (S-type innovation).

P-Type Innovations
With P-type innovation, an organization successfully builds a product that is innovative and triumphs in the

marketplace. The Polaroid story has P-type (product) innovation written all over it. In fact, I'm surprised that Edwin Land didn't get a Nobel Prize for his invention.

Another example of a P-type innovator, Pan American World Airways, was a global leader in product innovation. It was the first to create and utilize long-range weather forecasting, the first commercial airline to cross the Atlantic Ocean[3], and the first to transmit in-flight data to the ground.

S-Type Innovations

S-type innovators, on the other hand, blaze trails through their strategy, forward thinking and agility, rather than through their products.

Domino's Pizza[4] is an example of an S-type innovator that didn't change the product—just the way customers interacted with it. The company's original business model was that of a standard pizza delivery chain; customers called in their orders over the phone and could then choose to pick up or have their order delivered. Now, Domino's offers an app that not only allows you to place your order online but also shows you the status of your order in real time, from the time your order is placed until your pizza arrives. The company even developed a bot that can assist with placing an order and integrates with several apps and AI platforms.[5]

Essentially, Domino's transformed itself into an e-commerce company. And, moving into the future, the company is looking to utilize technology like robots and driverless cars to automate pizza deliveries, too. The question is, with the same product set, will it be able to survive for long?

When Organizations Fail to Innovate

P-type and S-type innovations both have their own place and time. TFOs should focus on both product-type and strategic-type innovations at all times, as they are often complementary. When organizations fail to do this, even dominant players in the marketplace can meet their demise.

When digital cameras arrived in the 1980s, Canon, Nikon, Fuji, and other manufacturers released their own versions of the digital camera. Polaroid, however, didn't release its own digital cameras until 1996, more than a decade later. By 2001, Polaroid had filed for bankruptcy.

What on earth happened to Polaroid?

If Polaroid had missed out on digital technology, it would be a different story. But, in reality, the company's ex-CEO and then head of research was the man who single-handedly convinced the president of the United States to apply digital technologies to military satellites. This happened long before any of Polaroid's competitors started working with digital cameras.

In 1976, the US Air Force launched the first digital satellite. It was a massive success and a huge step forward in the field of photography. Dr. Land wasn't at all surprised. He believed in the technology. In 1988, CIA Director William Webster even went so far as to declare, "The contributions Dr. Land has made to national security are innumerable, and the influence he has had on our present intelligence capabilities is unequalled."[6]

Dr. Land and his management team were obviously in a position to be the pioneers of the digital camera business. But they decided not to pursue it.

In hindsight, this decision seems totally irrational. But Polaroid made more money on instant-print cartridges than it made on the cameras themselves. Digital cameras did not have cartridges. Without the revenue and profit from selling cartridges, Dr. Land and his management team didn't think money could be made on digital cameras. They did not see the potential for a completely new revenue stream, nor could they identify the S-type (strategy) innovation that was needed to support a new product line.

Dr. Edwin Land's contributions to science paved the way for multiple scientific and technological breakthroughs in his time—including digital photography. But when photography began shifting from film to digital, the Polaroid company failed to innovate. As a result, other forward-thinking companies surpassed it. A new P-type innovation disrupted its predecessor.

Today, some of the forward-thinking organizations leading the disruptions are Google, Amazon, Microsoft, Lyft, Uber, and similar entities. On the other hand, doctors and healthcare specialists, nonprofits, social services, heavy industry companies, and similar entities identify themselves more as people-centric than technology-based. They choose to find their bliss in ignorance.

Amazon is a fantastic example of a TFO that has continually reimagined its role in the industry. What began as an online bookseller has morphed into a retail and entertainment powerhouse through the Prime service and an AI giant through the ground-breaking Alexa assistant. At the time of this writing, in the midst of the current COVID-19 pandemic, many thoughts are being expressed about how companies will evolve and who will emerge on the other side. But while the media are full of news of layoffs and organizations shutting down, Amazon is hiring upwards of 100,000 workers. Although Amazon's supply chain is still dependent on people and is affected the same as everyone else, I wouldn't be surprised to see self-driving delivery trucks rolling out soon, as well as advanced scheduling technology, which will overcome the dependency on humans even further. Amazon is already setting itself up to emerge as an even stronger organization.

Facebook is another strong example of a TFO. What began as a social network offered exclusively to college students has evolved into a big data giant. It has even crossed into cryptocurrency—something early adopters (and probably the company leaders, too) could have never imagined at the outset. Uber went from an exclusive ride-sharing company to adding a food delivery service. And Sunoco is a traditional oil and gas company, but now it is automating.

Following the successes of the organizations listed above, many large traditional organizations are coming out and declaring themselves as digital organizations or TFOs. Some of us are thrilled to see practically all major organizations turning into TFOs.

The primary focus of a TFO is **the constant reimagining of the company's space** and learning to attain revenue and business in ways its leaders never thought of before. In contrast, some organizations aren't willing to evolve—like Polaroid—and they get left behind.

Blockbuster is another prime example of what can happen when an organization fails to innovate. Blockbuster was obsessed with its current business model. It had strong future projections based on past performance. But there was a fundamental flaw in the projections: they didn't think anyone would ever stop renting movies.

Fast-forward to today. The movie rental business has gone practically extinct, thanks to on-demand video streaming services. Netflix has dominated this market for years, but now, there are several new competitors. Other streaming services like Prime Video, Hulu, and Disney+ have entered the market. So how will Netflix evolve? It will have to reimagine its role in the on-demand video space, because disruption is coming. The good news is that Netflix is doing well in reimagining the business. (We will talk more about this in Chapter 3.) Blockbuster, with its dominant position in the video rental market, could easily have become Netflix, but it refused to evolve.

According to Ampere Analysis, Netflix's global market share of streaming video subscriptions dropped from 91 percent in 2007 to 19 percent in 2019. With 167 million subscribers—60 million in the United States—Netflix has still maintained its lead position among competitors.[7]

Another example of a company that refused to evolve is Nokia. According to Safi Bahcall's book *Loonshots*, an internal group at Nokia pitched the idea of a smartphone (as well as an app store) in 2004. Management quashed the idea because they felt the investment was risky and couldn't guarantee returns. Three years later, Apple introduced the iPhone and took the market by storm.[8]

Misconceptions about Innovation

The idea of innovation is a hard pill to swallow for leaders who can project the near future in their organizations. If I'm running a company, I probably see the revenue forecast clearly for the next five years. I'd have to be crazy to make big changes, right?

It's a massive risk to step outside what's already working for your organization and do something innovative and different. But it's an even bigger risk to miss being first to market with an innovative idea.

Being first to market is incredibly important. To be first to market, **you must solve a problem before the marketplace even knows it's a problem**. But then, you'll need to continue evolving. We know that when you're first to market, you'll be copied very quickly. It's all about leveraging the lead, capturing market share, and developing intimacy with your customers. Most important, your entire organization must shift to a forward-thinking, TFO culture.

I was talking to the CEO of a major Canadian airline company recently, and he referred to his organization as "a technology company that commutes people in the air." The leaders of most financial organizations act the same way, with technology at their core. There are many similar examples of leaders from various industries publicly declaring such statements. While this group is excited and encouraged by the growth potential during these times of massive change, they may not be sure how to be successful and maintain that success.

As mentioned in the Introduction, society at large is worried about the future in light of massive, technology-driven transformation. It is unhelpful that the ongoing media narrative paints a picture of drastic change and widespread job elimination due to technology. This obviously encourages a state of hesitation and uncertainty.

For a long time, I believed that, when it came to technology, all individuals could be sorted into one of three Digital Competency categories, and that these categories defined their concerns regarding the future role of technology:

> ↳ Digital Natives (DN): Folks who have grown up with smartphones and can literally survive on a smartphone for days. (Like my six-year-old, who knows what food to order and what games to play. He can survive on a smartphone.)

> ↳ Digital Immigrants (DI): This group encompasses millions of people in authoritative and traditional leadership positions across the globe—including me—who have learned about and adopted technology since the beginning of their careers.

↳ Digital Visitors (DV): People who are reluctant to adopt technology for work or business but are aware of industry changes. A member of this group might have a smartphone and iPad but may only use it as a camera or to video-chat with their children or grandchildren.

However, I was wrong.

Based on deeper analysis, it became clear that regardless of their tribe—DN, DI, or DV—people in every category experience discomfort and pessimism surrounding the digital evolution. That means we have an obligation to demystify the digital revolution.

How Today's Organizations Respond to Technology

Organizations demonstrate varying comfort levels in this technology-centric world. Globally, they tend to fall into three categories:

↳ Technology Does Not Matter (TDNM): Organizations that have operated long-term without technology at their center and can't see the disruption coming to their industry. They assume they're safe to continue status-quo operations without responding to any massive digital changes on the horizon.

↳ Technology Scratcher (TS): Organizations that scratch the surface of technology without diving in completely. They fear missing out on new tech tools and want to be seen as innovative, but tech is still not part of their organizational DNA.

↳ Technology-First Organization (TFO): Organizations that are committed to making technology the center of their organization and part of their DNA.

Industrial Revolution 4.0 is the biggest change in the history of humankind. It's no wonder, then, that organizations exist on such a wide spectrum. It's also no surprise that there is still so much fear of technology.

Why are leaders so hesitant and fearful when it comes to becoming more technology-centric? I believe our brains, in a biological sense, have a big role to play in this phenomenon, in terms of how we perceive and handle changes around us.

The Brain's Response to Change

Our brains, for ages, have been trained to accept and engage with certain pieces of news and data while rejecting others. Good news isn't considered real news; bad news is more engaging.

Take, for example, the following two headlines:

↳ "Quarter over quarter, the employment rate has increased by 1%" (followed by a series of data by sector)

↳ "Another hate-crime in a bar full of people in the United States, 7 killed"

Our brain, by design, will remember and engage more with the latter piece of news. This creates a vicious circle. The actors—such as journalists, politicians, etc.—who bring news to most of our population are humans, too. Their key objective is to engage more with the population. As a result, the feed of increasingly dramatic news keeps coming, and our brain doesn't get a break.

From an evolutionary standpoint, our brains are shock-driven and wired to protect us from danger. Our fight-or-flight response was designed to keep us safe from predators. Now, we're in a world where we don't have to fight for food and protection the way our distant ancestors did.

The book *Factfulness* by Hans Rosling and Anna Rosling Rönnlund[9] essentially states that **the world is in the best state it's ever been in**. The authors argue that, worldwide, metrics are better than ever. Economic conditions, overall human health and the advancement of technology are all looking positive. I tend to agree. So, if the world is better than ever before, why do we hear dramatic or tragic news *every time* we turn on the TV or the radio?

Despite an objectively positive outlook, **change triggers fear and a fight-or-flight response**. News outlets and entertainment media home in on this, triggering fear over and over with shocking, dramatic headlines that highlight the worst world events and the ugliest aspects of the human condition. Often, we instinctively accept the shocking news, even if it isn't accurate. That's because the purpose of many media is to increase audience—not to inform. Reports on mass murders, tragic predictions, and fear-mongering are the order of the day.

How we respond to a specific piece of news depends on the facts provided and what our brains decipher as dramatic and shocking. Good reports of a decrease in unemployment are overshadowed by tragic stories of death and dismemberment. And when good stories are glossed over in the news, we don't absorb them. Instead, we fixate on dramatic negativity. We reject one type of news and accept the other.

The other day, my six-year-old son and I went for ice cream after dinner. When we arrived at the strip mall, there were police cars in the parking lot. My son immediately asked, "Daddy, what if somebody is here with a gun and shoots us when we get out of the car?"

My son's question paints a sad picture of our modern outlook on the world. He was simply repeating what we've heard on the news over and over. The media make it sound as though everyone is shooting everybody, but Calgary, Alberta, is a very peaceful city. Our reality is vastly different from the media portrayal of modern life, but that doesn't stop us from becoming fearful. Perhaps it's time to retrain our brains to take in information differently.

As previously stated, the human brain is a product of millions of years of evolution and is hard-wired with instincts that helped our ancestors survive in small groups of hunters and gatherers. We often jump to swift conclusions without much thinking—conclusions that used to help our ancestors avoid immediate dangers but do not do us any good now. Even if, on paper, the new Industrial Revolution has exponentially increased the rate of human evolution, our brains still follow primitive, linear-thinking instincts we inherited from our primitive ancestors.

We have many instincts that were useful thousands of years ago, but not today. For example, in ancient times, gossip and dramatic stories told by word-of-mouth were the only sources of news and useful information. Modern humans have access to television, the internet, and other media that provide us with information and entertainment—yet we are still interested in gossip and dramatic stories.

Similarly, our bodies still crave carbohydrates and fatty foods because they were once life-saving sources of energy when food was scarce. Sugar and fat are scientifically proven to lead to obesity, and today's doctors advocate limiting them in our diets. Yet our quick-thinking "modern" brains still crave these foods.

We still need these dramatic instincts or illusions to give meaning to our world and "get us through the day." If we shifted every input and analyzed every decision rationally, a normal life would be impossible.

Our brains need to shift gears toward the facts and realities of today. There are thousands of facts and figures being fed to the human brain throughout the day, thanks to media and digital platforms. I won't ask you to memorize a string of global statistics, but I do suggest that you learn what details to pay attention to and which to ignore.

All this is to say that **context is critical**. This is as true for organizations as it is for individuals. Life is a mixture of the dramatic and the mundane. While it's not necessary to switch off all media, we must all be conscious and discerning about what we're taking in. Pick and choose, learn what to keep and what to give up. By all means, use media for entertainment, but train your brain to look at the facts.

Recent Examples of the Modern Brain's Outdated Thought Patterns

Let's take a recent example from *Factfulness*. Democrats and Republicans in the United States often blame each other, each side claiming that the other doesn't know the correct facts.

A recent survey asking what percentage of the world's population is living in extreme poverty revealed surprising results. Only 5 percent of respondents in the United States answered correctly, regardless of their allegiance to Republicans or Democrats. The rest of the participants, regardless of their political preferences, believed either that the extreme poverty level had not changed or had doubled over the past twenty years. According to Rosling, the poverty level has actually decreased by half in the past twenty years.[10] It seems clear that you don't get facts when listening to the media.

We look at life through a North American lens, believing that we're the best continent in the world. We see other countries and people as odd, poor, or in need of our help. In reality, the world outside of North America has truly evolved.

Again, the facts are often different than our perceptions. In a study outlined in *Factfulness*, people in the US were asked about the child vaccination rate to assess their knowledge of basic modern healthcare across the world. Some people who thought themselves socially aware actually scored miserably on the survey. Only 13 percent of the respondents got the answer correct: 80 percent of children worldwide are vaccinated. The respondents mistakenly believed that a lower percentage of children were vaccinated worldwide, thanks to the many stories we are bombarded with about vaccination controversies.

This is indeed a case of mass ignorance—and don't for a second think that educated people get all facts correct! Highly educated people who take an interest in world events, such as scientists and even professors, also gave answers that were wrong. Our collective mental wiring goes deeper than just the "common human."

But it isn't just ignorance or a problem of upgrading our knowledge. **Our issues cannot be fixed simply by providing clearer data** through lectures and presentations. That's because we have an overly dramatic worldview, and our brains are hostage to our dramatic instincts. Our overly dramatic worldview draws people to the most extreme and most negative answers.

People intuitively refer to their worldview when thinking, guessing, or learning about the world. So, if your worldview is incorrect, you will automatically draw incorrect conclusions. This isn't a result of outdated knowledge; even people with updated information get it wrong. Rather, it is because of how our brains work.

By extension, when media and sources are talking about **the dramatic impact of the digital revolution**, it is hard to be excited.

I believe that when it comes to the most valuable asset available to organizations today, data is second only to people. In future chapters, I will outline data that will help us focus on the positive potential of the digital revolution. In addition, I will suggest some simple but extremely powerful habits to help you see the potential for an exponentially positive outcome. (More on that later.)

The Facts: The Promise of Technology and AI

The world has about 8 billion people in it today. The global population was about 1.6 billion in 1900 and 6 billion in 2000. It took thousands of years to get to 1.6 billion, yet it took only 100 years to quadruple that number.

Interestingly, over this period of population growth, the average number of offspring per woman has shrunk from about 5 to 2.5, so the growth is supported more by science than pure biological math. Science is a huge contributor. Overall health and economic conditions have improved drastically, which has resulted in higher survival rates and numerous diseases and other causes for mortality being reduced or eradicated.

The scientific and technological advances of the industrial revolutions of the past have gotten us here. Looking forward, the projections estimate that we'll be at 10 billion by 2050, and we'll flatten out at 11 billion by 2100.

We will be welcoming about 30 percent more humans into this world in the next 100 years. Into a world that is already significantly different from what it was a century ago. Into a world where personalization is a normal expectation. Into a world where the average number of offspring per woman will reduce even further—to fewer than two. Technology and AI will be an integral part of the journey. Technology and AI will have to sustain the population by providing food and identifying personalized needs and wants.

Today, more than 2.9 billion people around the globe are connected online. Within the next six to eight years, that number is expected to increase to nearly 8 billion, with every individual on the planet having access to a megabit-per-second connection or better. And, with the commoditization of AI, we are already on the verge of becoming a multiplanetary species.[11]

Really think about that: in a few years' time, *every single human* will have access to the world's information at multi-megabit connectivity and massive computational power on the cloud. And, with our widespread access to 5G bandwidth, the internet is even more efficient and more powerful than ever. [12]

Digital advancement is facilitating our exponential evolution. For example, a multitude of labs and entrepreneurs are working to create lasting, high-bandwidth connections between the digital world and the human neocortex of the brain. As Ray Kurzweil predicts[13], we shall see a human-cloud connection by the mid-2030s. Entrepreneurs like Bryan Johnson and his company, Kernel, are committing hundreds of millions of dollars toward this vision.

The end results of connecting your neocortex with the cloud are twofold:
- ↳ To increase the memory capacity of the brain
- ↳ To connect the brain to anyone else's brain via a global mesh network

Joel Garreau's book *Radical Evolution* takes this idea a step further, positing that technological advances are rapidly opening doors that could give humans veritable superpowers through the power of technology and AI.

Whatever progress we make, human evolution cannot overlook the benefits of the digital revolution, which is acting as a catalyst in the exciting evolutionary history of mankind.[14]

The Vital Role of the Digital Leader

People flock to specialists and experts in their fields because they're the most knowledgeable. When someone wins a Nobel Prize, it's not because a certain number of people report to that person; it's because of the knowledge he or she has brought to the world.

As I stated in the Introduction of this book, I don't consider the word *leader* to necessarily refer to the management of people. You might be a production manager or a leader at a production facility with *no one* reporting to you. Maybe your entire operation is automated, or perhaps there are one or two people who are present to assist you administratively.

A leader doesn't have to be a supervisor in traditional terms; a leader can simply drive a project or

advance transformation in their workplace. You can be a transformational leader regardless of who does or does not report to you.

In the book *Turn the Ship Around!* author L. David Marquet recounts a story about a naval submarine in crisis—not because it was under enemy attack, but because morale and performance were at an all-time low.

The men and women aboard the USS *Santa Fe* were followers, not leaders. When Marquet, the ship's captain, unwittingly gave them an impossible order, no one balked; instead, they tried to follow it without question. It became clear to Marquet that the traditional chain of command wasn't working for his crew, and this situation was the trigger. Marquet began advocating for a leadership model that expects and enables leadership at all levels.[15]

Marquet's model focuses on improving competence and clarity across the organization. It began to push authority closer to the activity. This new leadership model, when implemented in the sub, boosted crew morale and, essentially, turned the ship around, leading the USS *Santa Fe* to become an award-winning submarine that promoted a high ratio of officers to command positions.

Being willing to evolve and lead in this era of rapidly advancing technology is essential for anyone who wishes to become a digital leader.

For individuals who are feeling excited or nervous about becoming digital leaders, it's vital to follow four basic habits, which will be discussed in detail in Part 2 of this book. Embracing these practices will make you the digital leader everyone will wish to follow:

1. Be Authentic
2. Embrace Technology
3. Be Human-centric
4. Be Agile

The reality is, a very small portion of academia and niche organizations with large research and development (R&D) investments will keep evolving core technologies. However, most organizations and industries will be sourcing, implementing, and applying technology to their businesses. As a result, people and their Leadership 4.0 capabilities will be the key differentiators for most organizations. In the next chapter, we'll take a look at what exactly Industrial Revolution 4.0 is, and how it has affected—and will continue to affect—individuals and organizations around the world.

Key Takeaways

- Our brains often operate based on illusions and habits formed over thousands of years.
- Feeding facts and altering beliefs are two of the best things we can offer our brains to adapt to this exponential change.
- Despite some wrinkles, such as environmental factors, the world is currently at its best ever for humans.
- The rate of change happening around us is massively faster (probably ten times faster) and bigger than any of the past industrial revolutions, let alone overall human evolution.
- World population will increase to 11 billion from 8 billion currently.
- Almost 8 billion people will be connected through social media in six to eight years.

"Things are often much better than we think."

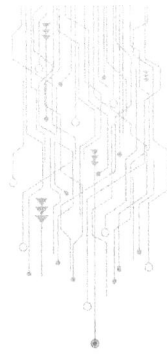

Chapter 2:
Industrial Revolution 4.0

In 2019, I visited the first Amazon Go Store near down-town Seattle. If you're a member, you scan your card at the gate, and it automatically opens the door. You go in, select items from the shelves, put them in your bag, and walk out. When you leave, you swipe your card and immediately get a bill in your email.

I experimented a few times by messing around with some items—moving them from one shelf to the other—but every time, the bill was accurate. How does such a system work? Overhead cameras monitor every movement and track codes for each item. You don't need to wait in a queue or even carry cash or credit cards in a Go Store! You can just walk in and do your thing.

Retail businesses such as Walmart and Target, all over North America and other parts of the world, employ millions of people either at their checkout counters or as store managers or executives. But with technologies like those used in the Go Store, all these people are set to lose their jobs, as they will surely be taken over by high-tech machines.

Imagine if you had a Go Store grocery market and a traditional grocery in your neighbourhood in the midst of a global pandemic like COVID-19. Where would you go? Convenience, coupled with the benefits of the safety of staff and customers, will almost certainly contribute to the future success of these stores.

Well, as it turns out, retail workers aren't the only ones whose jobs are at risk.

Financial advisors, insurance agents, cashiers, waiters, security guards, factory workers, and doctors: This is a shout-out to you. **You might all become redundant within the next fifteen years.**

Don't think I am exaggerating. In 2013, in a study titled "The Future of Employment," Oxford researchers Carl Benedikt Frey and Michael A. Osborne surveyed the likelihood of various professions being taken over by computer algorithms within the next 20 years.[16] The algorithm they developed estimated that 47 percent of jobs in the US are at high risk.

By 2033[17], human telemarketers and insurance underwriters could lose their jobs; the same might happen to sports referees, cashiers, waiters, tour guides, bakers, bus drivers, construction workers, security guards, sailors, and veterinary assistants as Industrial Revolution 4.0 affects the global job sector and invades the delivery of our daily requirements through AI, robots, and gadgets.

The physical world is becoming more and more entwined with the virtual world. The social media revolution embodied by Facebook, Twitter and Tencent has given everyone a voice and a way to communicate instantly across the planet. Today, more than 30 percent of people around the world use social media to communicate and stay on top of world events, turning the world into a global village.

A Preview of What's Coming

I began Chapter One with an insightful quote from economist Klaus Schwab. Schwab's point on the future of AI and its effect on humans bears repeating here.

The twenty-first century might see a drop in the value of humans who lose their economic and military usefulness as machines and AI take over high-precision work.[18] The system will find value in humans working collectively, but there might not be the same level of importance for humans as individuals. There are several patterns emerging from the incorporation of AI into the workforce in the coming decades, including the vulnerability of certain "routine" jobs and "predictable physical and cognitive tasks" to replacement.[19] Yuval Harari painted the picture very well in his book *Homos Deus*, which explains how the "great decoupling" will change the usefulness of humans as individuals.[20]

Humans will also lose their economic and military usefulness. Gone are the days of mass world wars. Armies of the twenty-first century are much more cutting-edge technology-driven, which requires small, highly specialized armies who can operate pilotless planes, fly drones, or wage cyber warfare. The warfare of the future may only last a few minutes, which may place a higher value on AI solutions over men and women.

AI solutions will be relentless, 24/7, data-driven—and fast. Things won't be much different when it comes to industry. Here's a quick rundown of what we have to look forward to:

↳ AI and data solutions will change the landscape and role of humans not only in terms of *what* the industry will focus on but *how* it will be delivered.

↳ Corporations will have to change their linear focus from "take, make, and waste" to a "cradle to cradle," sustainability-conscious approach.

↳ Environmental and social considerations will become just as important as profit in business.

↳ AI will play a big part in developing a sustainable future, which will be based on data.

We can also expect to see a major disruption in the way we understand politics. Although general elections will still be important, data, intelligence, and social media will heavily influence our political choices. Political leaders will be more focused on waging data-driven campaigns, which technology will help to personalize for individuals. In other words, AI will use the right election promises for respective groups and demographics. Campaign managers may soon become AI-based robots, able to do the work more accurately, more efficiently, and with a wider reach, 24 hours a day, seven days a week.

If you find it difficult to wrap your head around some of these scenarios, you're not alone. But it's better to face these changes head-on and try to understand them, as opposed to collectively ignoring them.

The Rise of AI in Product and Service-Based Businesses

While Amazon's Go Store may be mind-boggling, the company's online shopping experience is no less impressive. Amazon employs algorithms that constantly study you and then use the accumulated information to recommend products. When I go to a physical bookstore, I wander the shelves or ask the person on the floor for help choosing my books after browsing. When I am buying on Amazon.ca, however, the recommendation is given by algorithms that help me decide more quickly—and potentially buy more items than I had originally planned to purchase, as is often the case.

But, as mentioned at the beginning of this chapter, it isn't just the retail sector that will be disrupted. The medical sector may soon follow. A few months back, when I was travelling around Chicago and was a bit under the weather, I found an intriguing app on my phone. I signed up, put my credit card on it, and pressed the button. Before I knew it, a real medical doctor appeared on the video!

I shared my symptoms with the doctor via video chat. He asked me some basic questions and checked my throat using my phone's camera. Then he sent a prescription to a pharmacy near my hotel. I went there and picked up the medication. Simple, easy, effective, and all on an app. However, even this is bound to change. I believe that the role the doctor played in this case can be easily replaced by an AI program recommending medicine after mapping my symptoms and comparing the data to information known about common diseases.[21]

A human doctor usually recognizes a patient's emotional state by analyzing external signals such as facial expressions and tone of voice. Surprisingly, it has been found that IBM's Watson, an AI platform that manages complex data, can similarly analyze external signals—and do so more accurately than a human doctor. Beyond that, Watson could simultaneously analyze numerous internal indicators that are normally hidden from a human doctor's diagnosis. By monitoring several pieces of biometric data, the robot could even determine how the patient feels without being told!

Let's take this a step further and address the question of bedside manner. Thanks to statistics learned from several previous social encounters, Watson could determine what a patient needed to hear and tell them in just the right tone of voice. In the case of human doctors, their own emotions can be counterproductive. After all, the human brain can succumb to the emotions of the person he or she is talking to, but a robot will never have such emotions, thus leading to an objective response.

Although a host of technical problems still prevent Watson and its ilk from replacing most human doctors tomorrow morning, solving these technical problems is just a matter of time. The process might be complete in a decade or so, paving the way to an age when you might need just one Dr. Watson for *all* of your illnesses! And think of the financial implications. It might require billions of investment dollars to make Watson a "Super Doctor," but, in the long run, it will prove to be cheaper than training human doctors every year.

AI and the Fight Against COVID-19

As mentioned earlier in this book, the COVID-19 global pandemic is at its height at the time of this writing, and technology has empowered our worldwide response in a remarkable way. That said, there are also concerns about the ramifications of these changes and what our future might look like.

I have the utmost respect for the doctors and health-care staff on the front lines in our war against the COVID-19 virus. It takes a lot of heart, soul and courage to place yourself in such a vulnerable position. (I certainly have more respect and admiration for our healthcare workers than NHL players at this time.) However, in the long term, this is a fight between microbes and technology, and AI is one of our most valuable weapons.

The major factors that determine the success of any AI solution are the quality and the quantity of available data. In the future, other viruses may strike with similarly devastating results. It's all about how quickly it can be predicted, how effectively it can be diagnosed, how rapidly we can come up with treatment and/or a vaccine, and, most important, how we can connect and collaborate without political or geographical boundaries. Technology is the key to that journey. I don't think we're there yet. But even now, AI is playing a significant role—and a far greater one than most of us might guess.

On December 31, 2019, a Canadian-based AI system known as BlueDot predicted the COVID-19 outbreak over a week before the World Health Organization released its official warning on January 9, 2020.[22] Not only was BlueDot able to out-predict human researchers, but it also pinpointed the top twenty most susceptible cities, based on travel patterns from Wuhan, China.

AI can be utilized to effectively predict an individual's overall likelihood of surviving a coronavirus infection. Not only that: studies have already shown that AI systems can diagnose COVID-19 just as accurately as human healthcare specialists by utilizing chest X-rays.[23]

When it comes to mitigating the effects of COVID-19, a great deal of attention has been placed on "flattening the curve." As is to be expected, tracking the disease's location on the epidemiological curve requires a great deal of data from public health authorities. Fortunately, technology is being utilized to track and to predict the spread of the disease, allowing local healthcare systems to plan accordingly. Carnegie Mellon University, for example, adapted algorithms originally designed to predict the spread of the seasonal flu for use in the fight against COVID-19.[24]

While all this is quite positive, there is also a darker side to AI integration, and that has to do with its potential use in social control. According to an article in the *South China Morning Post, infrared cameras have been utilized to scan crowds in China and identify individuals with high body temperatures.* The infrared cameras in question can reportedly scan the body temperatures of up to 200 people per minute with just a 0.5-degree Celsius margin of error. But what's most startling is that this same system "*is also being used to ensure citizens obey self-quarantine orders. According to reports, individuals who flouted the order and left home would get a call from the authorities, presumably after being tracked by the facial recognition system.*"[25]

If this sounds like something out of science fiction, it's time to wake up to reality. In the United States, social distancing software is being developed to help ensure that safe practices are being observed, and authorities in Moscow, Russia, have unveiled a smartphone app to track people violating stay-at-home orders.[26] After all, mobile devices with AI-powered apps already regularly harvest information such as their users' demographics and health data—and, in many cases, their geographical location. Google has even published a series of COVID-19 Community Mobility Reports based on this data, showing the effects of reduced mobility on populations worldwide. It's not hard to see how this data could be used to identify high-risk individuals or even employed to enforce stay-at-home orders.

One also cannot help but wonder about the lasting effects this event could have on data privacy. As Yuval Noah Harari stated in a recent article, "*Even when infections from coronavirus are down to zero, some data-hungry governments could argue they need to keep the biometric surveillance systems in place because they fear a second*

wave of coronavirus, or because there is a new Ebola strain
evolving in central Africa, or because ... you get the idea.
A big battle has been raging in recent years over our
privacy. The coronavirus crisis could be the battle's tip-
ping point."[27]

I believe technology will play a much bigger role in
healthcare in the days to come, and the roles of professionals will be more focused on frontline execution.

AI in the Financial Sector

Even financial sectors will be disrupted by technology.
Many new banks and technology-based financial institutions have shown up in the past ten years, including
fintechs—computer programs designed to provide financial services. Fintechs do work similar to brick-and-mortar
banks, such as providing credit cards, approving loans,
and offering other financial services, but with far less
overhead. As a result of this smaller overhead, these digitally born fintechs provide more modern services and
platforms to their customers, and at a lower cost.

While fintechs are disrupting financial sectors,
even bigger disruptions to the financial sector are
coming from the "big techs." These are the large,
multibillion-dollar, data-rich technology organizations
like Apple, Google and Amazon.

Big techs have few major advantages over traditional banks. They are digitally born and hence nimbler
and more modern. They already have huge captive
customer bases, and they have more data about us—
more than our parents do. Among Siri, Google Assistant and Alexa, the big techs are privy to many of our

conversations and actions at home or at work. Apple Pay can now replace your credit card. Amazon knows my purchasing habits better than anyone else in the world. And most of my day-to-day purchases are made on Amazon anyway. Big techs are the biggest disruptors for the banks and most financial institutions, particularly in the retail space. There is a huge opportunity for day-to-day consumption, personal technology, and banking to converge in the near future.

On the investment side, most modern financial trading is already being managed by computer algorithms that can process data much faster than a human can blink. Trained AI can work on my portfolio, consuming petabytes of data around the clock, and provide me with recommendations and even do the trading for me.

Financial planning and portfolio management are becoming cheaper, faster, and more accurate. The algorithm is king—less so the human financial advisor. It's a trend that's here to stay, and the algorithms will only become more accurate. One interesting issue that remains unanswered is how one algorithm may differentiate from another in terms of intelligence.

Professions on the Fast Track to Extinction

Numerous professions are significantly changing or quickly becoming extinct. Even just a decade ago, they were completely secure from automation, but today, they're endangered species!

There are already businesses where, if you visit their loading docks, you will find no drivers. Instead, you'll find a small fleet of auto-driven cars, with a single person sitting there, overseeing all activities. Robots and 3D printers are replacing manual labour jobs such as clothing manufacturing, and highly intelligent algorithms will do the same to white-collar jobs such as bank clerks and travel agents. Think about it. Do you physically visit a travel agent when you want to buy airline tickets or book a hotel? It's easier than ever to take care of services like travel booking through an app or website with just a few clicks—no human intermediary required.

Lawyers might get a break, too. They may no longer need to consult loads of law books to help a client. Even the ability to tell if a client is telling the truth can be solved via simple algorithms, which may take over a large part of law-related professions.

Believe it or not, teachers—the first knowledge-givers for our children—could also become redundant. Scary, isn't it? The next generation may begin their learning process in virtual classrooms. Organizations like Mindojo are developing interactive algorithms that will not only teach math, physics and history, but will simultaneously study the behaviour and nature of the student.

Digital teachers will closely monitor answers given by students and examine how long they took to give that answer. Over time, they will determine where a student's strengths and weaknesses lie. With this information, they will be able to teach subjects that will best suit the personality of the particular student! (And, mind you, digital teachers will never lose their cool or punish a student.

But, then again, they also will never show love or affection to a first-time learner, who probably needs that human touch to fully embrace the love of learning.)

At the time of this writing, we are witnessing an example. Teachers in schools and universities have changed their role to deliver education using technology from their homes, and it is interesting to observe how everyone is adapting. My six-year-old seems to be into it. He just did a show-and-tell presentation on stock markets using Google Docs and sharing over Zoom. I especially love their physical education class, which is basically an online cardio class. Parents are even allowed to join in, which is extra fun.

In this example, the traditional role of teachers has completely changed. The many other jobs that are associated with maintaining and running a physical school building or university campus are not required. Your technology infrastructure *becomes* your building or campus. On the positive side, maybe this will bring down the costs of traditional education and lead to lower student debt.

The Downsides to Tech Evolution

While job loss is a scary enough scenario for many of us, there is a darker side to digital evolution, and that is its potential to make us vulnerable.

In 2013, Syrian hackers broke into Associated Press's official Twitter account. At 1:07 p.m., the hackers sent out the following tweet using the Associated Press's

Twitter handle: "Breaking: Two Explosions in the White House and Barack Obama is injured."[28]

The tweet was fraudulent, but within seconds of this tweet being sent to the Associated Press's almost 2 million followers, automated trade algorithms, which constantly monitor news feeds, reacted by selling stocks like mad. By 1:08 p.m., just one minute later, the Dow Jones went into freefall.

The Associated Press took only three minutes to correct the error, reporting at 1:10 p.m. that the tweet had been fraudulent.[29] But in just three minutes, $136 billion of market equity was effectively erased.[30] Fortunately, the Dow had returned to normal by 1:13 p.m.—five minutes after the disaster began.

In 2017, it was revealed that Russian hackers were penetrating power company grids across the United States.[31] The US government didn't respond to these allegations until 2018, and official governmental and organizational responses to this infiltration are still pending.[32] This could have disastrous implications for power in the US in the future if proper security measures aren't put in place.[33]

While these scenarios are certainly frightening, the twenty-first century may be poised for an unprecedented phenomenon: the rise of a massive new non-working class. This "new class" will not be merely unemployed but *unemployable*.

Humanity's New Class

Let's get into some statistics.

In 2010, only 2 percent of Americans worked in agriculture and 20 percent worked in industry, while 78 percent worked as teachers, doctors, web page designers, and so on. And the shift continues to the services sector: by 2018, the number of Americans working in direct on-farm jobs was around 2.6 million, or only 1.3 percent of total U.S. employment.[34] With the arrival of algorithms able to teach, diagnose, and design, humans will soon lose their significance in many services-sector jobs.[35] Often this is due to efficiency and speed, but sometimes it's because an AI-driven solution can process more information, do so more quickly, and provide better and more useful services.

Many new professions may appear in the future, such as virtual world designers. Other fields with potential growth include environmental sustainability and mental health, which will be the two biggest challenges for the next decade. We are still in the early stages of both of these fields.

Moving forward, society as a whole will probably shift its focus to the "how" just as much as the "what"—making an organization's process just as important as its outcome. Organizations will have to rethink the fundamentals of their businesses in the context of the planet and society. To drive this massive journey of intelligent technology, several new subject areas will emerge. These will include legal and ethical considerations for AI, but such professions will require much more creativity and flexibility than typical jobs of the past.

The question facing us now is whether a 40-year-old insurance agent will be able to reinvent himself as a virtual world designer. At present, we don't really know how such a drastic shift will play out.

The Threat of Social Inequality

The crucial problem won't just be creating new jobs but **creating jobs that humans can perform better than algorithms**.

With the triumph of algorithms over human jobs, power will be concentrated in the hands of a tiny group of elites who will own the all-powerful algorithms. An unprecedented social and political inequality might thus arise. The backlash seems inevitable. Those who have lost their jobs may form unions. They may organize strikes or create mass movements.

Alternatively, an age may soon dawn in which **algorithms themselves turn into owners**! Human laws already recognize intersubjective entities like corporations and nations as "legal persons." We may soon give similar legal status to algorithms.

We must not forget Klaus Schwab's golden words: "The Fourth Industrial Revolution can compromise humanity's traditional sources of meaning—work, community, family, and identity—or it can lift humanity to a new collective and moral consciousness based on a sense of shared destiny. The choice is ours."[36]

Enough shocking facts. I believe **the future will be amazing and full of new opportunities, but we will have to pivot to contribute**. For the right organizations and individuals, the opportunities are immense.

Busting the Media-Fueled AI Myth

A few weeks ago, a family friend who works in the oil and gas sector brought up the topic of the "impact of

AI" over dinner. He's an engineer and a transformation leader, driving change at work. As we went deeper into this topic, I could sense that he was scared about the impact of AI on jobs and society at large. I had to remind myself that my friend is a digital immigrant about four years away from his retirement. Although he is moving forward in his transformation journey at work, he surely doesn't *believe* in it. How can we drive real change with so many unsure digital immigrant leaders?

At the beginning of this chapter, I shared insights from Frey and Osborne's study, "The Future of Unemployment." Many people read the headline of this study and interpret it as doom and gloom. They see it as proof of the catastrophic threat to their livelihood.

Nobody reads the study—they just read the headline.

Reports like this are long and tedious, and they contain many caveats and assumptions. News outlets tailor stories and headlines to be short and juicy. And these days, when people gloss over these news stories on social media, the information is getting shorter and shorter. In turn, headlines like these lead to a large percentage of the world believing that technology will, indeed, spell catastrophe for the modern human.

"47% of jobs will be lost to technology in the next 20 years." That's the sort of headline we often see in the media.[37]

Really? I ask myself.

While I understand the data collected for these studies, I also can observe what's *actually* happening in the marketplace. I hear pessimistic statements every time a new technology is brought to market, but the reality is the economy is growing, and most global metrics are better than ever before. In short, I don't pay attention to such statements. I train my brain to follow the "reality data."

Misleading Job Loss Statistics

It's true that, as technology advances, the jobs we know and do today may disappear. But that doesn't mean that the jobs in question will be eliminated altogether, and it certainly doesn't mean these people will no longer be able to find employment.

Take the earlier Industrial Revolution, for example. An incredible 98 percent of textile manufacturing jobs in the United Kingdom were replaced by machines. But what happened next? People were hired to operate those machines. Textile costs fell, and demand rose. The market exploded, and job opportunities in that market rose dramatically.[38] The size of the pie, so to speak, grew.[39]

Recently, I was giving a talk to university students— my favourite digital natives—and I faced the same question over and over: "What will happen to everything we are learning at university or in the workplace if 47 percent of today's jobs no longer exist?" I felt that I had to address these students' concerns on the spot, because percentages are an interesting way to measure this kind of data.

A percentage is a great measure for some purposes but not so informative for others. Let's try to transform the percentage-based impact on jobs into absolute numbers. We'll need to make two assumptions for this.

Assumption 1: This change or impact on human jobs is already underway

Let's assume, for the sake of this discussion, that this impact began in 2013. (The media and several research groups have been talking about job impact for a while, so I am assuming 2013, as the Brookings Institute Study published in 2013 cited earlier in this chapter is one of the more credible research reports.)

Assumption 2: The number of jobs has a positive and direct correlation to gross world product

Gross world product (GWP) is the aggregate gross domestic product (GDP) of all the countries in the world. For simplicity in this discussion, I will assume that GWP is a proxy for job numbers. This is a reasonable assumption, since it is hard to determine the number of worldwide jobs accurately. (There is, after all, an abundance of unregistered jobs in the world.)

With the above assumptions, let's start a calculation on job impact:

↳ Gross World Product in 2013 = $77.19 trillion USD

↳ As per the statement, "47% of jobs will be lost to technology in the next 20 years" 47% of GWP in 2013 = 47% of $77.19 trillion USD = $36.27 trillion USD[40]

So, in 2033, $36.27 trillion USD of GWP will be taken in by technology instead of humans.

Let's now view what 2033 will look like from the GWP perspective.

Based on available forecasted data and historical trends[41], GWP for 2033 is expected to touch $150 trillion USD. From this $150 trillion USD GWP, if we take away the number of jobs affected by automation ($36.27 trillion as per the previous calculation), here is the result:

↳ Residual opportunity size = Net GWP (proxy for jobs) after impact from automation = $113.73 trillion USD (about 47% growth over 2013 GWP)

I personally believe the growth outlined above will be even greater than forecast. My advice: emerge from your media-driven shock and see things for what they really are.

We all need to stop behaving as if the sky is falling and robots will soon rule the world. **We have good reason to be optimistic,** because Industrial Revolution 4.0 is presenting opportunities to the world that have never existed before. Technology and AI are poised to grow the size of the pie, not shrink it.

Key Takeaways

- We are at a crossroads in an interesting era. This is an age when industry and society are both experiencing one of the most powerful revolutions of all time.
- While Industrial Revolution 4.0 has the potential to change the world positively, we have to be aware that technologies can have negative results if we don't consider how they can change us.
- We build what we value. This means we need to remember our values as we're building with these new technologies.
- The innovations in artificial intelligence, biotechnology, robotics, and other emerging technologies will redefine what it means to be human and how we engage with one another and the planet.
- In the coming decades, we must establish guardrails that keep the advances of the Fourth Industrial Revolution on track to benefit all of humanity.
- We must recognize and manage potential negative impacts, especially in the areas of equality, employment, privacy, and trust. This effort requires all stakeholders—governments, policymakers, international organizations, regulators, business organizations, academia, and civil society—to work together to steer powerful emerging technologies in ways that limit risk and

create a world that aligns with common goals for the future. Ethical innovation will be key.

- You—as a person, citizen, employee, investor, and social influencer—are a critical stakeholder in the Fourth Industrial Revolution. Sharing your thoughts on new technologies and what you value as this revolution unfolds is essential.

- The world we create through technologies will shape our lives, and it is this new world that we will pass on to the next generation.

- Let's focus on the overall size of the pie. The upsides for industry and society are far more numerous than the potential downsides.

"Focusing on the size of the pie will benefit the world and is the best recipe for mankind."

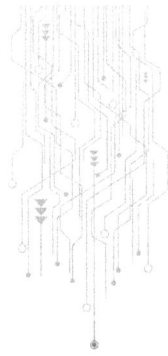

Chapter 3:

The Technology-First Organization

In Chapter 1, we outlined three categories when it comes to how organizations approach technology: technology does not matter (TDNM), technology scratcher (TS), and technology-first organizations (TFOs).

TDNM organizations consist mostly of organizations whose leaders believe they won't feel the impact of technological evolution. Nonprofits, healthcare organizations, doctors and other human-centric service businesses tend to fall into this category. Their top priority is not to adopt the latest tech; instead, it's to maintain the status quo under the assumption that the status quo will continue to work well. Their response tends to be, "Ultimately, we have to go and take care of people, and nothing's going to change that. The service we provide won't be so affected by technology."

Many of these organizations perceive technology as a tool that can only execute repetitive tasks that do not require much intelligence. They assume that human knowledge and the "human touch" are irreplaceable. Meanwhile, technology is evolving in terms of intelligent solutions and creating innovations, such as compassionate bots for real-time chats for mental health support and AI doctors for better and faster diagnosis of health conditions and diseases such as breast cancer.

Until recently, the resources industry didn't see itself as tech-centric. However, things are changing. Companies like Teck were not early adopters of technology, but now, they are very active in the digital space.[42] Technology is contributing to their bottom line.[43]

TS organizations tend to believe in the motto "crawl, walk, run." It is normal for a traditional organization to start out as a TS. In this world of exponential innovation and continuous evolution, TFOs like Lyft and Slack enter the market with great buzz, despite their meagre history of barely positive cash flow.[44] However, larger TS organizations are unsure of the path ahead.

It seems the stories of TS haven't changed much over the years. Typical organizations apply technology on an ad hoc basis, either to automate something they already do or to make it faster, cheaper, or better. In the past, organizations would create a portal or a website to allow their customers to reach out—ensuring convenience for customers and potentially lower costs.

Ultimately, the growth of TS organizations tends to be linear, as these organizations are focused on approaches utilized in the past. There's nothing wrong with continuous improvements, but in the Leadership 4.0 world, we are seeing organizations come up with products and services that did not exist as recently as last year. New products and services result in new revenue streams, redefining entire businesses, and TS organizations are not often redefining their purpose and adopting a technology-based vision.

Innovation Defined

Innovation may be defined or interpreted differently by different people. The mindset of most traditional organizations is that innovation equals continuous development. It's considered a victory if an organization improves by 10 or 20 percent per year. That's kind of the maximum.

What innovation means to me is continuously redefining products and services; that's P-type innovation. S-type innovation, or strategic innovation, is where an organization is doing something strategically different—redefining their revenue models from the ground up.

Many forward-thinking companies offer services that are totally free to use, which would have been a foreign concept as recently as a decade ago. For example, you don't pay to log in to Facebook or Gmail. Instead, their revenue models are always innovating, bringing in money from different sources. That's the fundamental difference between how traditional organizations think about innovation—continuous development—as opposed to continuous improvement.

I'm still connected with the friends I made during my time at Siemens, where I spent some productive years of my career. In 2014, Siemens shifted its purpose away from maximizing shareholder value to serving society through its technology-centric "Vision 2020."

According to CEO Barbara Humpton, "The biggest obstacle to any transformation is literally just the way we've always done things." Siemens' focus on the oil and gas business is being shifted toward endeavours such as digital industries, electric vehicle mobility, and energy efficiency. Similarly, China's AIA Group has moved beyond insurance to become a wellness company. Philips has also shifted away from the lighting business toward healthcare technology.[45]

While traditional organizations can only hope to improve by 5 percent or 10 percent, the organizations listed above have the ability to grow twofold, threefold, or even fourfold in just one to three years. These exponential organizations[46] (ExOs) are those who are agile in redefining themselves and their products and services, regularly.

Innovators are people and organizations who are continuously redefining their businesses. They aren't just improving; they are redefining their products, services, and strategic business models.

It's nice to see that a lot of major TS companies are catching up.[47] New technologies are helping in this regard because they're far more scalable than ever before.

Technology-First is More Than a Website

Twenty-five years ago, I started my first company with a project from a jewelry chain in Kolkata, India. Kolkata is a city of 16 million people and extreme competition. With no experience or contacts, it wasn't easy to get the first project. With a $200 advance for the work, I was excited and wanted to set up the team as soon as possible.

Frankly, I had no team or company when I got the project. I needed at least two more developers in addition to myself to achieve the committed timelines. I immediately hired my sister, who was a student at the time, and one of her classmates. I bought some hardware pieces and assembled and installed a Windows server, an access database, and a simple, traditional programming software: Visual Basic. We were ready!

Our client was the youngest adult member of a family that ran a famous jewelry business. In India, the jewelry industry can be a high-revenue business with high margins. Gold, silver, diamonds, and precious metals are considered investments, almost like stocks or futures—a hedge portfolio of investments in addition to being decorative luxury items.

My client wanted a solution to achieve the following:

↳ A system for consistent transaction management and accounting with the objective of eliminating inconsistency of practices across his store network and reducing the overall dependence on people.

↳ Complete and current customer information at a glance, with the objective of sending good wishes on special days, nudging customers with promotions, and providing special treatment for high-value customers.

↳ The possibility eventually for customers to directly transact and track their orders through the system. (This was achieved shortly after, when we moved the solution from Visual Basic to web applications written in Active Server Pages, JavaScript, etc.)

Every time I reflect on this first project of mine, it occurs to me that the work that "digital" groups are doing across many larger organizations isn't much different in principle.

Organizations that assume they are driving a major digital transformation by setting up website teams may be off-base. No doubt technology has upscaled and democratized, but fundamentally, DevOps—the methodology used in digital product development—and full-stack development were considered "business as usual" even 20 years ago, albeit without the modern acronyms and technology. Having a large omnichannel presence and being able to reach customers wherever they are is admirable. But if, as an organization, your aspiration is to become a TFO, the work doesn't stop at an e-commerce website or an app.

My client did well and, for a short period, disrupted his peer group, but he didn't turn into the Uber of the jewelry industry. Nor did my organization turn into Google.

My first company, 7P-Consultants, survived for about two years before I sold it to a Silicon Valley startup that turned it into their offshore centre. We had more business than we could deliver, and we were growing very fast. But at that time, India was evolving as an offshore destination for all major organizations of the world. As a result, I faced huge challenges in retaining talent; 7P had become an experience-gathering ground for fresh engineers who would soon move on to major corporations like IBM, Accenture, GE, etc. I didn't blame them. These were better brands with better overseas opportunities and benefits. Some of my former employ-ees are now seniors in many large organizations—including my sister.

The point is, an e-commerce website is a start, but it won't position your company on the leading edge of innovation. But continuous innovation and reimagining your place in the market will. A robust e-commerce plat-form has really helped organizations in terms of oper-ational sustainability during the COVID-19 pandemic. It has quickly become obvious that the integration of a robust supply chain was equally important, if not more important than ever. Organizations able to quickly make the shift in how they operate are emerging stronger.

How Can an Organization Become a TFO?

Digital transformation is larger than the digitization of activities. Organizations that are focused solely on that might find themselves wondering what's missing.

Being a TFO involves the following:

↳ Having **an active, overarching digital vision** infused in the company's purpose, which is com-posed of three perspectives:

- Re-envisioning the purpose toward the planet and society
- Re-envisioning the customer experience
- Re-envisioning operational processes (or com-bining the previous two approaches to re-envi-sion business models)

↳ Having **an auto-evolving organization compris-
ing digital leaders** who consciously transform the
business with the applications of leading technology solutions inspired by the vision.

- A traditional business strategy has a start date
  and an end date when you execute. But you
  must develop an organization and culture which
  sets itself up to adjust to new requirements,
  needs, asks and visions, because the goals are
  constantly changing.

↳ Having **digital leaders who are cognitively
invested and feel safe innovating** without
bureaucracy.

- Innovation, in this case, means the ability to
  continuously redefine your business product
  and services, strategy, revenue model, and business model—not just continuous improvement.
  The continuous re-envisioning of your business
  model, products and services is key.

Ultimately, an organization is made up of people. If
those people are digital leaders, the organization can be
a TFO. **With the help of the people on your team, the
autopilot evolution of your organization will keep
happening continually**.

Adapt or Die

As Industrial Revolution 4.0 is a reality today and is creating a disruptive impact on the very existence of many
enterprises, every organization now has to be a part of

this expanded digital revolution. The adoption of the digital way is no longer an option. Rather, every successful global organization will have to be a TFO to survive.

Consider Asian Paints, one of the largest paint companies in India and the third largest in Asia. Why do I choose this particular organization? Because it is a classic example of how a company that could have otherwise lost the race survived and grew by pushing an aggressive digital model.

Through their digital model, Asian Paints has been able to globalize and maintain decent growth—more than 15 percent annually for a decade—while increasing efficiency, transforming the customer experience and reducing environmental impact. Through digital paint mixing, it offers thousands of options for homeowners to turn their colour theme and dreams into reality.[48]

Building on this digital advantage, Asian Paints has successfully beaten competitors in the race and has been serving India's massive economy from the top. Asian Paints has now moved into 17 countries and continues to expand today. This would have been impossible if it hadn't strengthened its technology for manufacturing, order processing and supply chain. A good foundation alone isn't enough. Digital transformations set the stage for that foundation to grow. Even the company's website affirms that digital transformation will continue in the future, stating, "The road ahead is to integrate all our stakeholders including suppliers, employees, and customers and create an extended enterprise." Good words that every company should think about.

Asian Paints is on the right track. Personally, I've always loved their commercials. They talk about relationships, festivities, experiences, etc., and how their products and services make these more meaningful for customers. They are really gorgeous. (I encourage you to hit them up on YouTube.)

Another retail giant, Nike, has used digital technology successfully to transform the way it does business. We all know Nike founded the business on innovation; its products were "different," but still, it wouldn't have survived on its product and design alone. The company used digital technology to transform into a wellness company.[49] It even created a new business unit called Nike Digital Sport to build new digital wellness products and reimagine customer interaction through digital social media.[50]

This digital think tank somehow clicked. Nike was no longer a mere seller of sports products. It became an integral part of customers' lives. Nike continues to innovate as a fitness industry leader, fostering a wide ecosystem of experts in exercise, diet and mental health through its Nike Training Club app.[51]

Updating the Outdated Linear Business Model

In 1960, the average lifespan of an S&P 500 company was about sixty years. Today, that life expectancy is only twenty years, and it is expected to drop to twelve years by 2027.

A major factor in this drastic change is **technology-driven disruption**. Organizations that are leading in their digital journey and redefining themselves continuously will survive and grow. For example, IBM completely changed its portfolio by making a huge shift toward services, away from hardware and software. Amazon is constantly coming up with new and innovative service lines, and Netflix is another example, with 44 percent of its revenue now coming from content creation rather than content distribution (more on this later).

Meanwhile, many organizations are still living a paradox. While technology and the digital world are growing exponentially, the manner in which many large organizations operate and organize themselves is still based on linear models, hierarchical structures and traditional processes.

Value Creation

In recent years, new corporations have emerged via **value creation**. Startups have taught us how to quickly scale digital business models—even without major financial resources. Agile methodologies such as the Lean Startup Methodology[52] arose to boost the processes of product and solution development for customers. They develop fast. They measure fast. They learn fast. Organizations like Airbnb, Coast Bike Share and similar scooter-sharing systems are strong examples of this model in action.

In 2007, the eventual founders of Airbnb, Joe Gebbia and Brian Chesky, began renting out space in their apartment to travellers who needed a place to sleep. This developed into the business idea for Airbnb, which has exploded into a $38 billion giant.[53] These same startups are certainly not immune to failure, but even in failure, they have the benefit of speed, and they use error as a source of learning. They were born digital, with the DNA of agility and the scale of a digital software giant.

At some point, **your company will also need to go digital**. There are already close to $2 trillion USD in the hands of four digital giants—Apple, Google, Amazon and Facebook. So, think differently. Think about how you can accelerate your decision-making processes and how you can scale strategy, structures, processes and systems.

Establishing a TFO with strong digital capabilities makes new digital initiatives easier and less risky while also providing the leverage to generate new cash flow. On the other hand, synergies that result from strong digital leadership can free up cash flow for investment and help engage employees to identify new opportunities. Thus, these two components work in tandem, balancing each other and resulting in an ever-increasing digital advantage that can readily be put to optimal use. The net result? Excellence and maximum achievement.

How to Update Your Business Model to Be Technology-First

Updating your business model involves **continuously defining, redefining, and reimagining the products and services that you produce**. Continuously improving or automating a task—or many tasks—and making it cheaper or better would be an example of a traditional improvement. The strategic part of the innovation is how you sell, how your revenue model works, and how the costs change.

Nike reimagined how to connect with customers and athletes and became more of a wellness company than a shoe company. That's the kind of innovation I'm referring to when I talk about "redefining" products and services. Nike became a TFO, yet it all started with importing shoes from Japan.

As I mentioned earlier, Amazon started as a book company, basically just selling books online. But, as we all know, the company drastically redefined its role. Now, Amazon is about delivering services. It's about Prime and Alexa. It's about connecting with people. For the longest time, the company focused on revenue and market share—not on profitability. One letter from Jeff Bezos to the stakeholders clearly states that the company leaders are "obsessed" with their customer and are focused on revenue, not profitability.[54] And they're not done yet; Amazon is continually redefining and reimagining its role in the market.[55]

I was recently talking with the chief innovation officer of a large bank in Canada, and I asked, "Hey, what are your thoughts on fintechs? Is it a threat to your industry?" He just smiled and laughed. I pressed on, "You're not scared about fintech?"

"Well, no. I'm not, really," he said. "Fintechs have their place, but they are also dependent on us, the bank. I'm more worried about the big techs."

He meant companies like Amazon, Google, Apple and Microsoft.

I touched on this briefly in Chapter 2 during our discussion of the rise of AI. Many of us already use Apple Pay everywhere, and Google is testing the waters of the financial sector as well. All of these companies have the potential to swap their entire communities of technology users into financial clients. Banks are worried about big techs launching into the financial sector—or any business, really—because they have the data, the basic infrastructure, and the existing customer base to do it.

My favourite motto for larger organizations is "Think big, **act now**, start small, scale fast." It's my take on Jim Carroll's "Think Big, Start Small, Scale Fast!"[56] I like to add, "act now," which is an integral part that makes it feel complete to me. This is the way you can turn your linear curves exponentially and experience many more changes in a shorter period of time.

Brands that didn't exist ten years ago appear within the top ten brands now. We have ExOs running the show now that didn't even exist in the very recent past. The **evolution** and the **amount of change** (in respect to the ratio of time) has increased far more over the last several decades.

As we discussed in the previous chapter, ExOs are TFOs that have the ability to scale multiple times higher than their peers because of the use of new organizational models that leverage the technologies we have today to grow exponentially.[57] In the past, the average time required for a business to reach a turnover of over $1 billion USD was more than 20 years. Today, ExOs reach a turnover of $1 billion in just nine months.[58] These ExOs aren't theories. They exist in reality, and you see them all around you—from Airbnb to Google, Uber, and even Netflix.[59]

In 2013, Netflix CEO Reed Hastings announced that the company would transition from purely distributing content to producing new, original content. Hastings' memo read, "We don't and can't compete on breadth with Comcast, Sky, Amazon, Apple, Microsoft, Sony, or Google. For us to be hugely successful, we have to be a focused passion brand. Starbucks, not 7-Eleven. Southwest, not United. HBO, not Dish." Since then, about 44 percent of Netflix's business is content creation. This profound reinvention tripled the company's revenue, and its stock has increased at an annual rate of 57 percent (versus an 11 percent increase for the S&P 500). Its profits are thirty-two times what they were before this reinvention.

In these TFOs, most revenues and profits came from products and services that did not exist before. They focused on increasing the "size of the pie." Take, for example, Tesla, often considered the Apple of the auto industry. It has been revolutionizing the market with its electric cars. When Tesla announced that it would launch its new electric car model—the Tesla 3—at a price of $35,000 USD, the result was impressive. While the entire auto industry only sees a crisis ahead of it, Tesla achieved something rare. After three days, pre-orders reached 276,000.[60] These numbers were specifically from orders that required a $1,000 deposit (not full purchases). Within a week of the Model 3 announcement, Tesla received 325,000 reservations. This volume equated to more than $14 billion in implied future sales.[61] At the time, this was the largest-ever one-week launch of a product. All this happened with a very small marketing budget, riding on the digital car.

TED, the global platform of highly inspiring talks, is also an example of an ExO. With an investment close to zero, it has managed to scale a business model that made the company a global media brand—all this in just five years. For this, Chris Anderson, the creator of TED, made two highly disruptive moves: the lectures were available for free on the web (in a scenario where no company did this).[62] It enabled people to translate the lectures into their local languages, making them reach over 100 languages. And it created TEDx, which allowed scaling the model to different countries.[63]

Key Takeaways

- The digital revolution is here to stay. This is no dot-com boom waiting for a bust.
- Organizations have only two options: invest in digital and become a TFO, or get disrupted.
- Not every investment today will have a cogent business case, yet you need to make these investments and believe that they will pay rich dividends.
- ExOs are the result of massive transformative purpose and excited digital leaders who are cognitively invested in innovating.

"In work and life, purpose often represents your possibilities."

Chapter 4:
Exponential DNA

In *Exponential Organizations,* Salim Ismail describes why new organizations are ten times better, faster and cheaper than non-ExOs and what to do about it. This is a major lesson that bears repeating here. Ismail proposes that ExOs are the result of a simple formula called MTP, which stands for **massive transformative purpose**.[64]

We already know that humans usually engage and get motivated to do something if they're connected to a higher cause or purpose. But it seems that most organizations haven't yet learned what this connection means. Their missions and visions are words hanging on the wall, but words don't create a connection with people working in the company.

MTP is a unique super-message—quite aspirational and inspirational. It's like a single message that identifies and distinguishes one organization from all others. It reaches the hearts and minds of people and is primarily something declared with great **honesty**, **sincerity** and **trust**. All employees of Google know their mission is "to organize the world's information and make it universally accessible and useful." Tesla's mission is to "accelerate the world's transition to sustainable energy." TED champions "ideas worth spreading."

Some ExOs have unique visions and missions that drive them to excellence, but most employees of large organizations are unaware of their company's mission(s) because most organizations don't engage their employees in this bigger inspirational, aspirational mission. Hence, the employees fail to connect their work to customers and society as a whole.

There are two types of organizations: traditional organizations that engage in continuous **improvement**, and organizations that are continually trying to **innovate** when it comes to their products and services. Everybody improves on new technology, but there's a major difference between continuous improvement and innovation.

Traditional organizations have a long strategic plan, typically decided upon by the senior management—a handful of executives and leaders. The organizations then execute based on the strategic plan.

Executives, and sometimes those who report immediately to them, are usually excited about strategic plans and are aligned with them. But this alignment doesn't always carry over to the rest of the employees. It's difficult to get excited about something that you haven't contributed to creating and/or doesn't directly impact your personal success.

Some would argue that the execution of a strategic plan succeeds in proportion to the success of the organization and, hence, is connected to job stability and bonuses for employees at large. I don't disagree, but for most employees, it's difficult to see the correlation, especially when they're so out of the loop that they can't even articulate what the strategic plan is.

Depending on what kind of organization we work with, we spend our extra hour, or our discretionary hour, making sure that the boss is happy and that the key stakeholders are happy. We spend it that way because the motivation is to be at that level of comfort and to achieve the right number of bonuses, rewards, etc.

Instead of the scenario above, let's try to imagine a brand new startup, where an employee's salary and/or incentives and career advancement are tied directly to the product and services of the company. As an employee, you know you're really not going to get a lot of money until this company takes off and the product is successful. Naturally, you're going to put that extra hour into the product itself because, in this case, you see a direct correlation between your work and the organizational outcome.

Now, imagine a large organization that can operate like a startup. The employees are motivated not just by a plan that has been laid down by the people up top; they're involved in the whole redefining and reimagining process, and they are given incentive to do so. This is the sort of scenario that sets the stage for hyper-growth in an organization. That connection between work and results is an absolute must, but, then again, a message alone isn't enough to characterize an ExO. Follow-through and scaling are equally important.

Strategy is not a one-time activity. Even in a traditional context, follow-through and continuous updates to the strategy—as well as connecting and cascading to the employees—is important. But what we talked about in the previous paragraphs is even more vital. The imagination can't come solely from a handful of people; strategy should come from a large number of people in the organization who are motivated to execute on it. This is not something that happens naturally.

Motivation drives most of our behaviours as human beings. When cash bonuses are tied to the work that employees are "told" to do, employees will get good at executing on what they're told. On the other hand, when rewards and bonuses are tied to innovation, chances are that employees will connect with the purpose and contribute to reimagining the business. Organizations with the potential to grow exponentially bigger are those that have employees who feel proud to be part of that growth. The employees know that their hard work will benefit the entire organization in a positive way, and their motivation is no longer linked exclusively to money.

Enabling the right set of organization characteristics (digital DNA) allows an organization to survive the shift and follow through on their MTP. Fundamentally, integrating digital innovation into your business's day-to-day operations doesn't happen by accident—the people steering the ship need to be fully committed to the changes you're trying to implement and, most important, understand what they mean and how they fit into the overall strategy.

A New Kind of Leadership

The digital boom will also work on a business model of "network orchestrators" that rely on the network as capital, unlike traditional business models that rely on physical, human, or intellectual capital.

The network orchestrators model is a contributory model, with Facebook being a prime example. The model concentrates on meeting human needs beyond basics such as food, clothing and shelter. Instead, the focus is on self-expression, the democratization of information, ease of access, etc. This demands a new kind of leadership and workforce.

It is vital to get everyone excited about pushing your company in the direction of digital innovation. Major changes that can come from rethinking the core tenets of your business model can only be implemented through firm commitment and cognitive investment. It starts at the top, but it should belong to everyone. And, well, you cannot wait too long to modify or abandon business models. For example:

↪ The US Postal Service knew for years that its business model would unravel when email became ubiquitous.

↪ Kodak knew that digital photos would mean the death of film but lagged behind in its investment strategy and innovations to keep the company alive.

↪ Even though AOL knew dial-up subscriptions were falling, it still shied away from retooling its business model.

↳ In June 2017, Toys R Us[65] went bankrupt and was forced to close all of its stores in the US because the organization failed to make technological changes and get up to speed with other e-commerce competitors.

↳ In 1985, Steve Jobs[66] left Apple and went on to create NEXT, a computer company that was, in itself, a failure. The technology he created for NEXT went on to become Apple's Mac OS X and consequently saved the company. In essence, Steve Jobs left Apple, only to return and make it into the next Apple.

In all of these examples, timing was everything. **In digital transformation, time is king**.

Data—Your Second Most Important Asset

To quote Edward Deming, "In God we trust; all others bring data."[67]

It is essential to understand the importance of data. Thus, there are three clear capabilities into which firms should invest time and energy.

Furthest Imaginable State

To define the furthest imaginable state (FIS), an organization must identify the use cases and associated value propositions, then work backward to identify the data elements needed to drive this value. This calls for strong outcome-oriented and value-based mindsets, as well as robust data science capability.

FIS matters because there are organizations that are currently becoming digital, meaning they are embracing technology and redefining their business. If that isn't you, there's a high chance that you will be disrupted; it's a matter of either getting disrupted or being part of the disruption.

You want to reduce or mitigate the risk of being disrupted. If you are part of the disruption, you can grow or do something different with a great purpose—something you couldn't have done in the past. If you do it right, your chances of coming out on top are pretty high.

Data Elements

Second, all digital leaders across business functions, including IT divisions, need to recognize the need for— as well as know how to extract, transform, and use—the right data elements from internal and external sources for meaningful analyses.

Ongoing Data Governance

Finally, robust master data and transactional data governance are needed on an ongoing basis. This will ensure that structured and unstructured data can be extracted and analyzed in real time to drive superior business decisions. This is an interactive cycle that, if embraced correctly, will help drive significant value. (Starbucks invested in this cycle a few years back and is now able to reap the benefits of having moved from user segmentation to the aspirational "segment of one").

Data is the key, and it is the starting point for most innovation and AI initiatives. If your data asset isn't in good shape, it will limit the art of what is possible and will have a negative impact on your journey or your reimagining efforts.

Essentially, the value starts with useful visualization. Though visualization is not the FIS—and neither is it considered "cool"—I can't emphasize enough that it is a very promising starting point.

This book is being published during unprecedented times, during which the tracking and forecasting of COVID-19 has reinforced my belief in the importance of data-driven visualization, in this case to track the spread of this global pandemic. A study published by Towards Data Science and mentioned earlier in this book also referenced a list of forecasting dashboards ranked by *MIT Technology Review.* The highest ranked dashboards were those published by UpCode, Nextrain, and Johns Hopkins University. Also on the list were forecasting dashboards published by Microsoft Bing's AI tracker, the BBC, Healthmap, and *The New York Times,* all of which provide a global overview of data on the COVID-19 pandemic. The study goes on to share that "an increasing number of countries already have their own dashboards in place; for instance, South Africa established the COVID-19 ZA South Africa Dashboard, which is maintained by the Data Science for Social Impact Research Group at the University of Pretoria. ... And Tirthajyoti Sarkar has published a Python script to illustrate how one could extract data from *The New York Times'* COVID-19 dataset to create data visualizations of the progression of the infection."[68]

This is a perfect example not only of the integral role of technology in tracking and forecasting world events but the importance of visualizing that data to determine our response. What our global economy will look like on the other side of the COVID-19 global pandemic remains to be seen, but it's clear that technology will play a more vital part than ever as we press forward.

Combining Digital and Physical Commerce

A new term is currently making the rounds in the digital world: "phygital."[69] It refers to a marriage of the best of the digital and physical worlds of commerce. Virtual technology is fast merging with the physical world, making "phygital" the latest buzzword. Apple Watch epitomizes all that is good about this phrase.

Ironically, the less-glamorous, physical nuts and bolts of supply chains and logistics are extremely important factors driving the success of digital. It is quite interesting that one of the most critical capabilities to win in the digital world involves mastering the physical world. Think about milk, for example: the challenges of predicting volumes required by various geographic locations and the infrastructure required to ship and store such a perishable commodity, with marginal incremental costs. Hence, winners in the digital world will be those companies that master the physical world.

The Rise of Phygital Stores and Commerce

We're living in the digital era, and it's no longer deemed bizarre to buy all kinds of products over the internet. In fact, one study found that 70 percent of internet users make purchases online.[70] Yet many of these same consumers prefer the experience of a physical interaction at a brick-and-mortar store. And still others browse for products in physical stores, only to ultimately make their purchase online. It's a clear demonstration of consumers' need for **immediacy**, **immersion**, *and* **interaction**.[71]

The market has responded to these trends with the rise of "phygital," which incorporates the best of both worlds—aspects from physical and digital sales environments—in an attempt to create a new kind of customer experience.

Phygital blurs the line between physical and digital by utilizing aspects of digital commerce within physical stores.

In Chapter 2, I talked about the Amazon Go Store, which forgoes the traditional checkout counter for a more digital experience: browse, scan, walk out, and receive an electronic receipt later. This is a prime example of the phygital consumer experience at work.

Even fast-food brands are getting in on the action. It's not at all uncommon to find touchscreens in fast-food restaurants where customers can place their orders. In China, KFC has taken this a step further, utilizing facial recognition technology and artificial intelligence to offer personalized orders for their customers.

The trend toward phygital shows no sign of slowing down. For millennials and Generation Z consumers, phygital feels like a more natural shopping experience that can deliver on their needs for immediacy and immersion, without losing the all-important element of personal interaction.

Purposeful Acquisitions and Strategic Partnerships

Major organizations have just two options in the Digital Age. They can grow, or they can get left behind. It's not difficult to see why Amazon, Facebook, Google and Microsoft all have a similar mindset when it comes to competition: If you can't beat it, buy it.

While some companies are busy attempting to create their own versions of popular services, a company like Amazon opts to look at what already works and then buy it. Amazon's noteworthy acquisitions include audiobook retailer Audible, video game streaming service Twitch, and IMDb. Amazon also famously bought organic supermarket chain Whole Foods Market in 2017 for $13.7 billion. In doing so, Amazon not only acquired strong businesses but a vast amount of consumer data to put to use in its existing business model.

For a while, Instagram was seen as a major competitor for Facebook. Facebook's response? They bought it. And that's just one of the 70-plus companies Facebook has acquired since 2004, including Oculus VR and

WhatsApp. Google, meanwhile, bought Motorola in 2012 and video hosting giant YouTube back in 2006, in addition to acquiring more than 200 other companies over the years. And Microsoft's list of acquisitions includes Hotmail, Skype and, more recently, LinkedIn, at a whopping $26.2 billion.

With all these purchases came a lot of data, which is how the major tech organizations got where they are today. But sometimes, it makes just as much sense to partner with another major organization to create a mutually beneficial arrangement.

Strategic partnerships often allow companies to expand their offerings without the need for expensive R&D. Examples that spring to mind include:

↳ A recent partnership between Microsoft and SAP aimed to ease SAP's customer journey from legacy systems to its new version, SAP S4, in the cloud

↳ Intel and Texas Instruments, together creating processors and computer chips for computer manufacturers

↳ Starbucks coffee shops in Barnes & Noble bookstores

↳ Windows Phones: the result of a partnership between Microsoft and Nokia

↳ Partnership on AI: an initiative featuring collaboration among Amazon, Facebook, Microsoft, IBM and other tech companies

↳ Nike+: a partnership between Nike and Apple

↳ Soundtrack for Your Ride: a partnership between Uber and Spotify

Strategic partnerships like these translate into consumer experiences and business values that otherwise wouldn't have been possible or would have required extensive R&D.

How to Get the Best Results with Digital Tools

Let me give you an example of how you can be smart and use digital tools to move in the right direction and get the best results.

Entravision Communication Corporation, a Spanish-language media company, operates more than a hundred radio stations, television stations and digital platforms. They also had a huge amount of data and realized they needed to employ this data through analytics to derive customer behavioural insights, which are, indeed, very important to organizations that sell products and services. The demand for deep insights into Latino markets began to grow, increasing the need for analytics and predictive modelling. The answer was Luminar, a business unit devoted to serving external clients through big data, created in 2012.[72] True to its digital vision, the company has gained clients across the spectrum, from Nestlé and General Mills to Target, among others.

The vision of this company shifted by using digitization in the right direction. The company had previously envisioned itself as a traditional broadcasting group and had experienced a slowdown in business as a result of the market's transformation. But it reinvented itself and adjusted quickly.

Luminar leveraged data to transform its business and protected itself from disruption, threats and risks to its business. It also seized the opportunity to innovate and develop emerging partnerships with complementary organizations.

We must remember, in this age of ground-breaking digital innovations that happen almost every day, **successful business models do not last forever**. As with Entravision Communication Corporation, venturing into uncharted territory may require creating a new value model.

Even competitive threats from other organizations in the same sector can act as a catalyst to start a new business model. New entrants may use the potential of digital technologies and can move ahead of you. Thus, you need to **act fast**.

Ask yourself some questions:

↳ Are you experiencing a gradual decline in traditional revenue streams or margin erosion due to commoditization?

↳ Are new competitors emerging from unexpected and adjacent territory?

↳ Are cheaper digital substitutes for your products or services making inroads in your market?

↳ Are traditional barriers to entry coming down in your industry?

Make agile decisions, follow flexible visions, and make maximum use of the digital vision.

The approach to this new Digital Age brings out your originality and keeps you authentic. **Re-envisioning the customer experience, operations and business models** is key to maintaining this approach.

Pharmaceutical giant Novartis, a traditional organization, actually changed the lives of customers, elevating them to a new high by re-envisioning its customer experience. CEO Joseph Jimenez wrote, "The technology we use in daily lives, such as smartphones and tablets, can make a real difference in helping patients manage their own health. We are exploring ways to use these tools to improve compliance rates and enable healthcare professionals to monitor patient progress remotely."[73]

Tencent Holdings is a Chinese tech and video game company that was founded in 1998. Its purpose was defined as "implementing our online lifestyle strategy, which strives to cater to the basic needs of our users." But in 2011, Tencent underwent a massive transformation under CEO Pony Ma Huateng. Its newly defined mission was "improving the quality of human life through digital innovation."

Online forums and video games had been at the core of Tencent's business model, but after adopting the new mission, it focused on fintech, educational content, entertainment and self-driving cars. Tencent was the first Asian company to surpass $500 billion in market valuation, thanks to this transformation, and it's worth noting that it continues to innovate: As of 2019, the company mission is "Tech for social good."

Attacking the "Bug"

One thing is clear: you need to transform to make it in the digital world. Every company has to join in this change and become a TFO. But this transformation won't be easy for traditional organizations.

One potential complication is that when you try to develop or implement a disruptive innovation in any large organization, its "immune system" will attack the "transformation bug" of the new system with all its strength. **Disruptive innovation will be seen as an intruder** attempting to break the existing system, bringing new ideas and breaking paradigms.

The truth is, this is a path of no return. Those who fit the best and anticipate changes will survive. Others will not. This is your only chance of survival, and, like Steve Jobs said, "You cannot connect the dots looking forward; you can only connect them looking backward. So you have to trust that the dots will somehow connect in the future."[74]

Imagine your future as a company of digital leaders working together in a TFO that has been newly reimagined for the Digital Age. If you're a company of 20,000 people, that means you could potentially be making use of 20,000 minds. Imagine creating a framework where 20,000 brains can be contributing to your organization's future. The fundamental factor is to imagine this scenario at play **throughout the organization**, versus a top-down strategic plan of execution.

Again, think of a small startup where everybody is motivated to create something different for the customer. In this situation, we are obsessed with the customer, because that's where our motivation is. That's the only path to success.

If you can somehow align the motivation of thousands of people—or hundreds of thousands of people working in a larger organization—so that their motivation is connected to the leader, while imagining your customer experience, then your customer obsession outcome will be different.

A new organization, or digital organizations that have the emerging capacity to increase, include everybody as they move forward. Everybody in the organization is motivated, and their motivation is aligned with this "reimagination" event.

Like any startup, digital organizations are motivated to redefine their success. They know that unless they do, they don't have a career. They don't have a future. They don't have financial stability, so they are motivated to reimagine the business and are absolutely obsessed with customer success.

Now, the question is, what are the positives in digitally built or digitally native companies? They are **continuously delivering products and services**, they **develop new models**, and they're **constantly changing their revenue and cost models**.

When you follow these steps, your revenue year over year increases multiple times—not by 5 percent or 10 percent, but exponentially. Things like that really do happen to organizations that continue reinventing themselves.

Digital leaders play a critical role in organizational innovation going forward. Technology has become a commodity, and people are the differentiators.

Key Takeaways

- Most organizations do not engage their employees in their larger inspirational and aspirational mission. Hence, the employees fail to connect their work to customers and society as a whole.
- Digital visions have three perspectives: re-envisioning the customer experience, re-envisioning operational processes, and combining the previous two approaches to re-envision business models.
- In the new digital world, you need emerging partnerships.

"Reimagining business is like making an original movie; it's different from a remake."

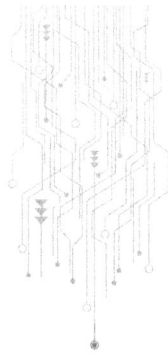

Chapter 5:

Technology Is a Commodity; People Are the Differentiators

As you move toward making your organization a TFO, remember that people are the key to a successful digital transformation. Technology is actually secondary to your team.

I like what the chief digital transformation officer of the world's largest enterprise resource planning (ERP) product once said. "Software is a tool. It is configurable," said Chakib Bouhdary, vice president, Value Engineering, at SAP America. "It all comes down to how it is being used and how you measure its value. We have seen the same software being used by two companies in the same industry. Some use it to their advantage. Some make a mess of it."[75]

I still have nightmares about some large ERP projects that left me with permanent scars, as well as a ton of experience. I also have several glorious success stories. In fact, I have so many stories about ERP that I could write another book on it. The net takeaway, though, is that **it was always the *people* who made such projects successful or made them failures**. In every case, the same technology solution was available for *all* organizations in the market, including our competitors. It's the other factors that make it or break it.

The "Solution Blocks" Approach

ERP projects are usually huge—typically multi-hundred-million-dollar projects—and include a lot of change within organizations and how they do their work. They also include a lot of change for the people within those organizations. And changing the core of how an organization traditionally conducts business is often very hard for people.

Most organizations have a lot of great people. They are mostly operations-minded. And when these big changes or projects come along, things can get difficult because a lot of change occurs in a short period of time. These initiatives create a great deal of stress and tension within organizations. Anyone who's been a part of major projects like this will second this comment. They are never very easy, and the transition is never smooth, and it tends to be a struggle to stay on schedule and on budget.

However, through dedicated leadership focus, I have also seen some large implementations proceed fairly well. When I look back, it was surely the **clarity**, **competence** and **courage** of the people involved that always made the difference.

You might assume that in fields such as data science and AI, things would be different—that technology is a differentiator. Since we're still in the earlier stages of the maturity lifecycle for products and solutions in the technology space, it sometimes feels that way. But I would argue that it is not. Rather, the key differentiator is the people involved.

In the 1980s and 1990s, everyone was busy building their own accounting, sales and marketing software, until enterprise resource planning (ERP) software and a few publicly available products standardized the space. In the same way, basic solution blocks in AI and data science will be commoditized and standardized, like LEGO bricks, over the next five to ten years. Technological breakthroughs will happen at the laboratories of major universities or research-based large organizations. The rest of us are essentially integrators, implementers and consumers of technology.

For the traditional functions in an organization like finance, accounting, HR, operations, etc., most companies will go and buy some kind of an ERP, cloud-based solution and implement it. Today, nobody would start a project to develop it from scratch in a customized manner for themselves. This was not the case in the pre-ERP age. Twenty or even fifteen years back, people would grapple with these processes and problems one at a time, organization by organization. Over time, these have been mostly standardized. We still implement some special solutions in our organizations or customize or specialize, but in most cases, the common belief is that most of the required functionality can be met by available solutions. (Examples include SAP, Oracle, Workday, and SuccessFactors.)

Over a period of time, the research part of this AI phase will be standardized and commoditized and made available to most organizations. And they will focus on the application and implementation part of it, as we do for ERP today. We don't develop SAP; we buy and then implement it.

Data science and AI are novelties at this time for most of us. As a result, we build solutions from the ground up, organization by organization and use case by use case. But over the next five to ten years, a similar evolution will happen. Research labs and large research-oriented companies will utilize what has been learned and encapsulate AI aspects in resuable, business-friendly applications and components. Just as with ERP, most of us will implement available solution components, building them together like LEGO bricks, to achieve the required business outcome. Value will come from the quality of data and the ability to imagine and implement solutions, instead of the capability to build from the ground up.

Even in the current state, success with AI is about digital vision, leader commitment, and an organization's ability to reimagine and implement solutions to correct business problems and take advantage of opportunities.

The Democratization of Technology

Today, the technology used across all organizations is pretty much the same or similar and is very easily accessed, with practically no entry barrier in terms of cost. Technology is truly being democratized and commoditized even as we speak.

Now and moving into the future, access to technology for you and for other companies will be the same. Individual organizations don't have to deal with the highly supercharged process of technology research. Instead, they have to focus more on the implementation and application. And that's where it becomes more of a people play; it's a matter of process, versus a matter of technology.

This is why having **the right people**—individuals who can reimagine your business, apply those "LEGO blocks," and make sure that your organization achieves what it wants and needs to achieve—is becoming far more critical than the technology itself. That's why I started with the example of ERP implementation. Most large organizations are on SAP or Oracle. What distinguishes them from one another is how those tools have been implemented and what else has been done or incorporated to run their business.

The bottom line is, **technology, in itself, is not a competitive advantage for your organization**. For the average organization, whether it is ERP or AI, no technology arbitrage exists. The differentiation exists in the digital vision and how your digital leaders are applying technology to achieve that vision. The traditional applications are now pretty much standardized to ERPs or point SaaS solutions. So nobody would say that one company is better than another because they have a certain ERP. It's hardly even a statement. Similarly, over time, AI will become even more standardized.[76]

Technology cannot distinguish you as an organization; how you implement and adopt it is the key. Thus, **the integration and usage of technology in an organization is a function of the quality and quantity of digital leaders in that organization**.

The differentiator is the people. *How* team members apply and use technology is what makes the difference—between themselves today and tomorrow, and between themselves and their competitors within the ecosystem.

People–The Key to Massive Transformation

In Chapter 1, I mentioned that data was the second most valuable asset of an organization. I believe that in this Digital Age, people are the most valuable asset.

The good news is, I haven't met any team member, leader, or CEO in recent times who isn't aware of the importance of digital transformation. Awareness isn't a problem anymore. Most have accepted the fact that they have to adapt to these digital changes for their success. However, there are many who are still skeptical: "Is this a thing for smaller companies and startups, or is it applicable to larger organizations as well?"

Those in larger organizations tend to ask, "Where do we start, and how do we begin the journey? What kind of organizational structure do we need to start this? How do I need to change myself? What do I need to do from the perspective of skill and acumen to become a successful professional in this era of digital revolution?"

In the previous chapter, we covered the first steps your organization needs to take in order to become a TFO: establishing an active, overarching, purposeful digital vision, setting your organization up to auto-evolve, and upgrading or hiring digital leaders to live out the organization's purpose. Digital leaders—humans—then lead the leaders and innovate to align with the organization's purpose. Purposeful digital vision is greatest when it's committed to and shared by the entire organization and with Wall Street.

More and more organizations are making public statements of massive digital investment and value. Auto-evolving organizations effectively do P-type and S-type innovation while effectively driving the current revenue cycle. Above all, **successful organizations consistently invest in building a population of digital leaders** and enable them to achieve the first two steps (a digital vision and auto-evolution).

Massive transformation is about defining a purpose that connects your organization with the planet, with society, and with the shareholders. And finally, it's about having the right people who can do that. For example, Google's stated purpose is "to organize the world's information and make it universally accessible and useful." A strong purpose gives your team something to rally around.

So, where do you start? First, upgrade and hire digital leaders who are excited to live out that purpose and empower those leaders to innovate in alignment with it.

There are questions galore and many theories about evolving into a technology-first organization, but the starting point for digital organizations is to increase the number of digital leaders throughout every level of your organization. When a business has a critical mass of digital leaders across different levels—not just at the top—it changes the culture, and the technology that is applied as a result helps to achieve TFO goals.

To recap, to turn your organization into a TFO, you need to:

↳ **Adopt a massive transformative purpose**
↳ **Upgrade and hire digital leaders who are excited to live out that purpose**
↳ **Empower your digital leaders to innovate in alignment with that purpose**

Digital Leadership Fundamentals

The success of an organization is almost directly correlated to its status as a TFO. And the probability of an organization turning into a TFO is directly proportional to the number of active digital leaders in the company. But what exactly is a digital leader? Let's take a moment to define the term.

Digital leaders are workers in an organization who "lead" the outcome, applying the fundamentals of digital leadership. A digital leader needs to be someone who is concerned with the customer/user experience of the services or products and the evolution of the revenue model, and who doesn't get hung up on functions that are still working. (They can't have an "if it's not broken, don't fix it" mentality.)

A digital leader needs to be comfortable not only embracing change but leading the charge. And, as we'll see in Chapter 8, digital leaders can't be afraid to roll up their sleeves and get their hands dirty, experimenting with innovative solutions that could help the organization achieve its purpose.

You need the hearts and minds of digital leaders cognitively invested 24/7. It may sound odd, but frankly, the nine-to-five work schedule and the concept of "work-life balance" don't exist for digital leaders. To be clear, I don't mean that digital leaders work longer hours; what I mean is that their work and life are integrated, and they own that identity all the time. **It's not about hours—it's about cognitive investment and commitment.**

Let's further clarify what digital leaders look like.

Habits of Digital Leadership

How do you know if someone is a digital leader? And how can you become a digital leader yourself? As previously mentioned, I have identified and mapped out four key habits of digital leadership:

1. **Be authentic**
2. **Embrace technology**
3. **Be human-centric**
4. **Be agile**

With relatively low costs and few barriers to entry, it has never been easier to access technology and become a TFO. The advent of the cloud has made it even more accessible. So, in order to become a TFO, where should you go from here?

It is important to treat employees well and develop a workforce that will act as key differentiators and add to the competitive advantage of an organization. A product or a marketing strategy can be replicated in time; however, a great set of digital leaders who stick by you and work toward a meaningful vision will lead to a great institution and higher profitability. Hence, it makes sense to invest in that all-important factor known as "your people."

Part 2 of this book will specifically describe the characteristics of a digital leader at the level of the individual. For digital natives and digital immigrants, I especially hope that this section helps you achieve success in Industrial Revolution 4.0.

Key Takeaways

- The same technology is used by most organizations. Some take advantage of it, and some make a mess.
- Technology arbitrage does not exist. It is all about the application and usage of technological opportunities available across different organizations.
- Digital awareness is present across organizations. What we need are committed digital leaders to take us through the digital transformation.
- Technology is democratized, and people are key to success.
- Digital leaders get to be at the front of the disruption and are not disrupted.
- A digital leader can be anyone in an organization, irrespective of title. Digital leadership is more about how we apply technology to transform business and less about how many people report to you.
- There is a convergence—organizations need digital leaders to become TFOs, and individuals, to be successful, need to level-up to become digital leaders.

_____

"Technology is as good as the people who implement it."

_____

Part 2:

Digital Leadership for Individuals

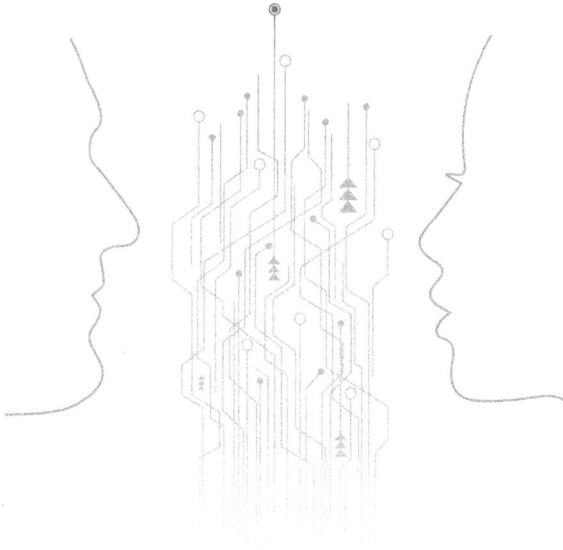

Chapter 6:
Authenticity

According to the 2018 Edelman Trust Barometer, the public's confidence in the traditional structures of American leadership is now fully undermined and has been replaced with a strong sense of fear, uncertainty and disillusionment.[77] The writing is on the wall. More so than ever before, we need authentic leaders who inspire trust, confidence and loyalty.

Authentic leaders will inspire trust and help pull everyone out of this slump by **demonstrating self-awareness, honesty, and courage**, by **building honest relationships based on their real values**, and by **listening to the people with whom they work**. Authentic leaders also aren't afraid to express themselves honestly, ask the difficult questions, and take action based on what they hear.

I've come across many authentic leaders who do not mind sharing their fears, likes, dislikes and their own leadership path with their teams, in a very open and honest way. Bill George, author of *True North: Discover Your*

Authentic Leadership, says, "Being authentic as a leader is hard work and takes years of experience in leadership roles. No one can be authentic without failures, saying and doing things they will come to regret. The key is to have the self-awareness to recognize these times and listen to close colleagues who point them out."[78]

Apply your knowledge to be original. The most thought-provoking leaders understand the skills, competencies and technical information needed to manage effectively in the modern workplace.

Leaders need to be authentic. These days, you're always connected. So is your family, and so are your friends and colleagues. Being authentic will help you foster an environment around you with a culture of acceptance of diverse thoughts and ideas—which, again, is crucial to innovation and transformation.

One of the best descriptions of authentic leadership comes from *Forbes* contributor Henna Inam, who writes, "Authentic leadership is the full expression of 'me' for the benefit of 'we.'"[79] Such leaders will reveal themselves in the way they think, speak and act, which builds trust among the team and followers.

Without trust, no collaboration or connectivity is possible. I ask of digital leaders: **Be yourself**, **connect individually**, and **walk the talk**.

Be Yourself

There is a lot of focus in every industry around diversity and inclusion (D&I). Every day, when you open social media, there are news stories such as, "GM now has a female CEO and CFO," or "Google has an immigrant with an accent as their CEO." After decades in which most industries were dominated by white men, we keep looking for visible minorities in the name of D&I. Many top tech organizations are now headed by senior leaders who are immigrants from Asia. We have made great strides toward including a wider variety of people whose voices can now be heard.

D&I, to me, is one of the biggest game-changers in working toward exponential transformation, because it brings out thoughts, ideas and perspectives in their fullest form. It allows more voices to be heard and applied. But the truth is, skin colour doesn't really matter. Accent doesn't matter, either. What matters is our ability to show up as who we really are, regardless of how we look or sound.

We can't change the world; what we can control is ourselves. We're probably harder on ourselves than others are on us, so this starts with showing compassion to ourselves.

To fully show up as who we are, we must first *accept* who we are. Prejudice still exists, but once you show up as yourself and accept yourself, you give the person with prejudice an opportunity to accept you as you are.

There is an interesting side benefit of being yourself. Stress is the single biggest challenge the workforce in North America faces today. Being yourself liberates a leader from that stress to some extent. In contrast, trying to emulate someone—to be something you're not—raises your stress levels.

D&I starts with being yourself. It depends heavily on demonstrating specific behaviours for others, and this starts with acting as your authentic self, showing yourself compassion, and showing up as the best version of you.

All digital leaders must strive to be themselves. As you bring your true self to life, a few things will happen:

↪ You bring your diverse thoughts to life.

↪ You make it psychologically safe for others around you to bring forward their true selves without distortion.

↪ You let go of the stress of being "someone else" at work or life.

↪ You enable everyone to express themselves in a safe environment, which is essential for innovation.

It's important for you to have an intimate sense of self-awareness so that you can truly be authentic. For example, I know myself very well and try to stay in tune with the things that work best for me in my interactions and practices as a digital leader.

I am a "highlight" reader. My attention span is shorter than that of many of my colleagues and team members. My fast reading speed is very helpful in taking quick actions and advancing the net conclusion, but it isn't a great attribute when details are important. I am open about it, and this helps me surround myself with readers who tend to be analytical and read in detail.

I'm also an early riser. I love my mornings; I can go anywhere, anytime early. But my energy and cognitive power go down in the evening. Instead of fighting this,

I accept it as being part of who I am, and I try to organize my day around it.

Authenticity in Action

Over the years, I have learned the importance of showing up as my authentic self. But I have also experienced the discomfort of trying to fit in with something that was against my wiring.

Having been part of the business development process in many of my past roles, there was a time when I had daily dinners and late-night sessions with my clients. I love meeting in-person with my clients. I get a lot of energy from those conversations, and I always come out happier on the other end, and with some new ideas. But I'm not much of a drinker. For a while, I assumed that I had to drink to fit in. I had a key customer executive who liked martinis with olives. Me, not so much, but I had martinis almost daily with her for years. Instead of politely declining and having my preferred green tea, I felt that I couldn't maintain the rapport of our meetings without having a martini, too. It was tiring to go along with the martinis—something I didn't truly want to do. I always left those meetings feeling tired.

I learned from those years. I still am proud of my commitment to customers and business, but I now believe that I could have handled it differently. Now, I let everyone know up front that I am a morning person and not someone who enjoys drinking past midnight. That doesn't mean I won't go out for occasional long dinners. But I often drink tea while everyone around me has expensive wine.

I used to struggle to display my true self. I don't hesitate to do that anymore. Most importantly, as a result, I believe I connect more authentically and have happier, more fulfilling interactions.

I am sure I am not alone. There are many leaders like me. When I'm myself, it does help others to feel safe and express themselves. Believe me, this works. And once you accept yourself, it opens the door to accepting others with open arms—which effectively improves D&I in your organization.

Once I was leading a merger and acquisition initiative that was very time-sensitive from the technology integration perspective. I was lucky enough to hand-pick my leadership team based on several factors. It was well known that this would be a high-speed, high-visibility, high-risk engagement. It was obvious there would be many long days and some weekend work.

For me, the team is the most important success factor most of the time, so I decided to take some time to talk and set expectations for each member of my integration team. One of the leaders was open with me upfront about his expectations. He basically said that he intended to maintain his work/life balance. He didn't appreciate the idea of long days or weekend work. He wanted me to know that he was committed and interested in the assignment, but he believed that he could get most of the work done within the timeframe of normal days.

It took a lot of authentic courage for this employee to be himself, and I really appreciated that. My respect for him as a professional actually went up. We experienced success, and he did very well overall. We were both respectful in our conversations and our social agreement.

Conference calls over Skype, Microsoft Teams and other internet-based software is one of the most common methods of communication in today's world. With D&I come differences in accents, which is an issue I have some experience with (even I have a fairly thick accent).

To be honest, earlier in my life, I sometimes avoided speaking up because of my accent. I now realize that, every time I gave in to that impulse, I killed an idea or two. It was not only a loss for me but also for the organizations I was working with. When the expected population growth of 2-billion-plus people occurs in Africa and Asia over the next couple of decades, leaders will have to become better at accepting and getting comfortable with various accents. I am extremely deliberate about making every conference call psychologically safe for the participants in all ways, regardless of the diversity of accents or ideas.

Be your best self!

Connect Individually

We like to be treated in a personalized manner, for who we are. So does everyone around us—our colleagues, our customers, our suppliers, etc. Today's technology makes this possible.

Our previous world was based on broad segmentation along the lines of age, income, geographic location, etc., but that's just not enough anymore. Technology and digital leadership are all about giving personalized and customized offerings to people. Take, for example, Netflix. We want top picks based on the kinds of movies and shows we like to watch. And any time you're shopping for something online by searching Google, your Facebook or social media will show ads for the items you just saw the next time you log in.

The individualization of technology has made it more focused than ever before; it's no longer about segmenting demographics of people.

With the possibilities opened up by technology, digital leaders are expected to make personalized connections with stakeholders around them. Connecting individually is a high-priority need, and it's important to cater to that need.

Every opportunity I get, I try to understand the people I work with. It is extremely powerful to connect with people on a personal level. It gives you context and helps you treat humans as the unique individuals they are—not as sales reps, developers, etc.

I worked for a highly intelligent man for a long time. He kept on asking for a dinner invitation at my place so that he could meet my family. I wasn't all that excited about the prospect, as we had just moved to a new place and were still settling in. I also had a small child who wasn't old enough for dinner conversations. But my boss literally *insisted* and came over with his girlfriend. It was not the fanciest dinner I'd ever hosted, but it helped me learn about the power of connection.

The key result was that we got to know each other's families, meaning we could talk about our loved ones by names rather than with pronouns. It made things much more real between us. Every conversation is more meaningful now. Today, he will randomly send me a text about his son winning a soccer medal, and I will send him pictures of my son with his chess trophy. This connects us as human beings. And when you connect as human beings, everything you do at work will become easier to share—and easier to disagree or agree with because you are connected individually.

To connect individually with the individuals on your team, you can follow easy, actionable steps, such as:

- ↪ Get to know your stakeholders' families and interests outside of work
- ↪ Stay connected with former employees and colleagues
- ↪ Talk about your shared interests
- ↪ Do not hesitate to be vulnerable
- ↪ Emphasize in-person connection
- ↪ Utilize social media to stay connected
- ↪ Believe that human connection is far more powerful than organizational lines

When you have a pool of smart people who are connected and personally committed, magic happens. Ultimately, human beings are social creatures and have high social needs. Work connection, augmented by personalized social connection, leads to the formation of a high-performing team, which is always a good thing.

I have worked for many leaders from different backgrounds, ethnicities and styles. I know their family members, and they are all my friends outside work. In fact, I struggle to understand how you could successfully work with someone without being emotionally connected. I still call my colleagues from twenty years back and stay connected. In many cases, I'm sure we will never cross paths again, so this isn't a way to keep in contact for future material gain. It's about the person. It just feels good!

I recently called one of my bosses from twenty-five years back on a Saturday morning. He's retired now and currently works as a professor. We talked about books (specifically the book *Homos Deus* by Yuval Noah Harari) and had a good chat. It was amazing. I certainly couldn't have bought that sort of pleasure with money or a fancy lunch.

I love every opportunity to meet people in person and get to know them, their families, and their aspirations. It helps me connect with their minds and souls. To be clear, an email won't cut it. I get so many emails every day, and I can't read all of them with the same level of attention. I tend not to feel as connected with email as I do over the phone or in person. There are many technical ways available these days to connect more periodically. I reconnected with my schoolmates after a gap of twenty years through Facebook. I have to admit, as much it costs us to give out our information to Facebook, it has definitely changed our lives.

Connecting Individually

There was a time when marketing books taught us to utilize a broad categorization of the market, grouping customers based on income, gender, etc. That format doesn't work anymore.

Organizations need to connect with all individuals, irrespective of their age, gender, skin colour, etc. The power of connecting individually is enormous. If you got on a plane and the cabin crew came up and greeted you by name, saying, "How are you doing? I haven't seen you for a long time!" that would just make your day, wouldn't it? It's simple but powerful. That sort of individual connection sweetens us up as customers, and we become more loyal and more willing to establish a long-term relationship with the brand because they know us and see us.

Digital leaders should connect individually with the ecosystem, their colleagues, their customers, and their employees. Great leaders are consistently self-aware, because they understand the impact they have on others. This helps them to always act according to their core values and, in turn, maintain integrity.

Most of us have personal narratives about defining moments. In our own hero stories, we surmount the obstacles along the way and learn invaluable lessons that shape our decision-making.

Think about times when you felt good about a meeting or an introduction to an individual. The other individual probably treated you with **attention** and **importance**. Odds are, you learned a few things and that person was probably in a "giving mode."

Giving is the most powerful thing in this world. Providing others with strong, personalized advice, providing a needed sense of connection and confidence, is one of the things that connect us all.

I remember being on a call with a CEO, along with one of my colleagues and other stakeholders. He made sure to take the opportunity to highlight us as the "finest of the company." I left the call feeling good, and I had learned a lot as well. The call left me with a stronger sense of accountability above and beyond my job.

I also recall meeting up with the ex-CEO of a large upstream oil and gas company and board member. What an amazing lady! It was meant to be a one-hour meeting, and my initial intent was to get some advice on board-level opportunities. The meeting went longer than an hour and forty-five minutes. It was my day off, so I wasn't in a rush, but for a first-time meeting, that's not normal.

I didn't leave the meeting with any board-level leads; instead, I left with something much more valuable. I got a deeper lesson on giving. Right after that, I read the book *Give and Take*.[80] I feel I became a better leader as a result. To me, that conversation was worth at least a year of executive education at a top university.

As a leader, don't miss out on any opportunity to give. It is amazing and leaves both sides feeling better.

David Marquet, author of the book *Turn the Ship Around!*, once mentioned using standard deviation to measure the success of communication in a meeting. I've decided to bring this concept to the home front, to take a look at some objective data.

> *My nephew (DN) comes home from school, his backpack on his shoulder and his phone in his hand. He is fourteen years old and a very bright and well-rounded human.*
>
> *Mom (DI): "You always have that phone in your hand. What do you do with it all the time? You just came back from school. At least take the time to tell*

me how your day went instead of looking at your phone. How was your day?"

Nephew: "Good."

Mom: "See? You don't talk to me anymore. You don't share anything. You are always on your phone. Here you go—supper is ready. What time is swimming practice today?"

Nephew: "Five."

Nephew leaves for swimming.

# of words used by the mom: 76

# of words used by my nephew: 2

Standard deviation of # of words used in the conversation:

$$=\sqrt{((76-39)^2+(2-39)^2)/(2-1)} = \mathbf{52.3}$$

That's very high. Obviously, a lower standard deviation would be better. It would be close to zero if Mom and Nephew spoke a similar number of words.

How can it get better? It can get better in two ways: either if my nephew used a higher number of words or his mother used fewer words. In particular, the words that were spoken *about* **my nephew** and not *to* **him** could easily be avoided. Maybe I am exaggerating to make the point.

How Can We Close the Gap between Digital Natives and Digital Immigrants?

1. Try to **get outside your comfort zone** in terms of the media you use for communication.
   - Maybe start texting/Snapchatting to a DN, or put some extra focus on a more traditional communication form.

2. Contribute to **minimizing the standard deviation**.
   - Communicate more *to* each other instead of *about* each other.

While connecting individually, a leader must remember the following:

↳ Maintain a harmonious work-life integration. How, when and whether to engage is a matter of personal choice.

↳ The manner in which we carry ourselves, the tools we use to communicate and connect, the ways we interact — all of these can make a world of difference to how effectively we can lead.

↳ It is hard work connecting with each individual to know their perspective and what they are best at, but this will help you to get the best from your team.

↳ Give at every opportunity you can find.

Connecting individually is the true manifestation of inclusion and encourages individuals to contribute their best. It also builds human connection overall, which lasts the longest.

"Work the Talk"

We are being watched every moment, not just the moments when we walk by.

Every time I, my team, or my family post something—or even like something—on social media, the ecosystem is receiving data to react or build an opinion. Thus, we have no choice but to do and behave the way we say. We've got to talk the talk *and* walk the walk.

The bottom line: Practise what you preach. You must do as you say and never say anything you don't truly believe. And never, *ever* say you'll do something you aren't really willing to do.

In the past, only organizations and institutions had "brands" because they spent money on advertisements in the media, deliberately creating the asset we know as a brand. But **brands are no longer just assets for organizations**. Courtesy of social media, we all have a visible, individual, personal brand these days.

Building a Personal Brand

Before interviewing anyone for a position, I always check the person out on LinkedIn. I remember someone in my team recommending a contract project manager who happened not to have a LinkedIn profile.

I have to be honest: I was baffled. Technically, having a LinkedIn profile isn't any sort of requirement for the job, but it has become a de facto standard for interviewers to search for one regardless.

Another favourite example of mine: when there's a material disconnect between someone's LinkedIn profile/resume and the facts they share during an interview. Often, I see candidates referring to their resume before answering every question. I find it ridiculous, then annoying, because it takes a few minutes to call some of their references and find out the truth. I'm not sure why candidates get into this kind of behaviour. It's a problem of overselling, which removes the first layer of trust before it is ever built.

Long-term, I have a feeling that this whole idea of submitting a resume for a job application will be a thing of the past. Google, Microsoft, Facebook, or one of the other big techs may soon provide screening services to the organizations doing the hiring, in exchange for fees. In many ways, this would be a much more accurate process anyway.

Be Authentic, Be Vulnerable

Thanks to social technologies, the world has become so much smaller and more visible. I rarely receive a profile for an interview that is beyond my second-level connections on LinkedIn, regardless of where the person is in the US, Canada, India, or any other part of the world.

The moral is, as you make connections online, be authentic. Be an individual, connect individually, then do as you say.

If your connections see a difference between what you say and do, they'll notice. They will see through the inauthenticity if, say, your LinkedIn profile presents a completely different persona from your Facebook profile.

Recently, I was co-hosting a town hall meeting over a Microsoft Teams broadcast with my leadership team. Even though we did a bit of a dry run, I still missed a slide. I guess I was too excited to share my opening remarks. Basically, I missed the introduction of the panel who was with me addressing the larger group. When I realized the mistake, instead of trying to hide it, I just stopped and said, "Hey, I made a mistake again. I do this all the time! I missed the introduction completely.

Let's go back three slides and do the introduction." It was easy, funny and authentic.

Expose your imperfections and be vulnerable. Trying to cover up a gaffe could lead to even bigger mistakes. And making excuses about those mistakes will make you seem inauthentic.

I can't help recalling Steve Harvey's misstep at the 2015 Miss Universe pageant, when he accidentally announced the wrong winner. After the show, he claimed the teleprompter had given him the incorrect information. His cue card, however, revealed the correct winner. Whatever the case, he might have been better off simply owning the mistake without trying to explain it away.

The most important thing is to be relatable in an authentic way. This feeling is liberating and rewarding.

At the same time, "be yourself" doesn't mean you stop developing yourself.

As Bill George says, "It really gets down to the lives you touch every day in your life ... and people you don't even know—sometimes whom you've impacted by who you are, what you stand for—by being true to what you believe."

Listen to yourself and listen to the people you lead or hope one day to lead.

⚱ Key Takeaways

- Don't feel like you can't admit it when you don't know something.
- Authentic leaders are all about asking questions, listening to the answers, and leveraging the strengths of those with whom they work.
- Be yourself, connect individually, and walk the talk—those are the three golden rules.
- Giving is the best thing you can do for yourself and your stakeholders.
- Giving is rewarding, and giving the right gift to the right person never reduces anything.

———————

"Being yourself is the best path to inclusion, and it results in a more engaged you and others."

———————

Chapter 7:
Embrace Technology

Technology is not optional, regardless of whether you are in human resources, sales, or a technology group. Every professional individual is expected to utilize tech at the very least.

Having an email address is no longer an option. Fifteen years ago, we could get away with saying we didn't have access to email over the weekend. Now, that's no longer possible. Like organizations, individuals are expected to be more tech-centric than ever. But the question remains of how much technology one applies and how they apply it to turn themselves into digital leaders.

The answer is: more than you think. Again, it's all about habits.

I was at a technology conference last year. Many highly intelligent speakers and panellists were present, but I especially loved the panel that was made up of three young ladies from universities, all in their early twenties. A senior technologist was interviewing these three DNs to get their viewpoint on various things in the future. Their responses were not only insightful but, at times, surprisingly different from those of DIs like me. It was refreshing that we were (finally) paying attention to the thoughts of DNs.

I learn so much from DNs thanks to their natural inclination toward technology. As a rule, they are significantly smarter than DIs like me, especially when it comes to embracing technology. The reality is that for the next two decades, DNs and DIs will have to work together, and technology will be what levels the playing field.

Perhaps, despite everything you've read so far, you still think your industry is exempt from needing to embrace technology. Well, a decade back, I led the IT group for a large railroad in Canada. One Sunday morning, I got a call about a train that was immobile. As a result, it was blocking traffic. Even the Royal Canadian Mounted Police got involved; it was a big deal!

My team was already involved in solving the problem. After some digging, it turned out that in one of the applications, the order in which records were being sorted was creating an issue. The solution to this catastrophe was a small piece of code in an SQL statement.

Think about that for a second. The train was invented over 200 years ago, but it has advanced with modern technology over the decades. Suddenly, a minor issue in the database can cause a major disruption for the railway. Clearly, information technology has become an intrinsic part of the operation. This scenario applies to every industry on the planet.

We are all aware of the potential for cyberattacks to shut down services in major healthcare systems, airlines, credit agencies, and so on. Whether we like it or not, we are strongly affected by these events, and we are all susceptible.

Technology is essential to our lives, which means we're all at risk. The possibility of a data breach or cyberattack just comes with the territory. An organization that neglects to protect itself in this regard is just asking for trouble.

The point is: Technology is no longer something that's "nice to have." It is at the core of all businesses. Hence, **digital leaders focus on technology as a priority**, as it is inseparable from the business.

"Learning and Unlearning" Technology

My son cannot comprehend how I could have organized a social life during the pre-Facebook era. He thinks I'm ancient! But why?

I had my first personal computer after I got my first contract for my startup 7P-Consultants. It was a mad dash to assemble my first workstation over an afternoon, and I had to buy some pizza for a friend of mine who knew hardware to help me get it all put together. Today, digital natives find it is hard to imagine how anyone could do their job without computers, email or mobile phones.

For my six-year-old digital native son, an iPad is a necessity of life. I can't even remember when he started using it, but I do remember that the first one he used smelled like yogurt and spilled milk all the time. (Trust me—we cleaned it pretty often.)

Theirs is a world of sitting on a couch with their iPads, Skyping their friends on the other side of the globe. By the time they become parents, their children will think that the technology they use now is as antique as the pagers and shoebox-sized "mobile phones" we used once upon a time. To keep up with these rapid changes in technology, digital leaders need to be open to the idea of **learning, unlearning, and relearning technology**.

Learning

Adult education experts estimate that around 40 percent of the technological know-how that we are learning today will be obsolete a decade from now. Indeed, the top ten most in-demand jobs today didn't even exist ten years ago. To say that we live in a changing world understates both the pace and the scope of ongoing change taking place in the digital world.

For the 3-billion-plus people in the workforce, it's not just about keeping up with the rate of change and the nature of the work we do, but how we do it and where we do it. When anyone can work from anywhere, it changes the nature of work everywhere. Traditional boundaries disappear, and the global talent pool becomes more skilled and mobile, which presents a challenge for people in developed countries to simply stay competitive. Our only choice is to adapt through the learning, unlearning, and relearning process. Here are a few tips to help you get started:

- ↪ **Learn something new** without trying to be overly safe, secure and comfortable.
- ↪ The time has come to **embrace new technological ideas with open arms**. Shed your rigidity in adapting and accepting the "new."
- ↪ Do your best to **mould the changes going around you to fit current "mental maps."** This applies to people, events and the general technological environment.
- ↪ **Embrace a product mindset**. New business models, technology and customer demands challenge leaders continuously.
- ↪ In the Digital Age, **there is no way back**; it is a one-way ticket.

Companies used to hire based on experience and skills. If this person knows a specific skill, let's hire him or her. But today, that's no longer the sole consideration, because what you know today might not be very relevant in the future. It's much more important that you have the basic core competency of **learning and relearning**. That's because the workplace isn't a linear journey toward a static destination. It is a constantly changing pathway.

To me, curiosity and creativity are becoming the two most important characteristics in the future.

There was a time when access to knowledge was costly, and it was only imparted at premium institutes. Now, enormous amounts of knowledge are available for free. Check out websites and offerings like www.edx.org or Coursera. High-quality learning can be obtained at zero cost. Today, instead of money, what you need to invest is time committed to deliberate learning.

My dad was a lifelong accountant at a mining company in India. He had an interesting theory. He used to say that relevant knowledge in any field increases by about 10 percent every year, and 10 percent of your acquired knowledge becomes irrelevant or is lost every year. That means there is a 20 percent depreciation of knowledge every year. In other words, if you don't upgrade yourself, your book value will be zero in just five years.

I'm not sure if Dad's theory makes sense from a mathematical standpoint, but I know for sure that it had a profound impact on me. I read one book a week. People who know me know that the best gift for me is a book.

Bill Gates reads about fifty books a year, which is about one a week. Gates has told reporters that when he reads nonfiction books, he tries to concentrate on the new knowledge he is acquiring through reading. His natural curiosity at times makes him sit with a pen and paper, jotting down the new things he learns in the process.

When I first heard that, years ago, I thought, "A book a week? I don't have time for that!" Then I thought, "Well, how does Bill Gates have time for this?" That doesn't make sense, does it? If Bill Gates, one of the busiest men in the world, has enough time to read a book a week, then so do I! I had to *make* time for this.

If you think you don't have time, you are not alone. You have to find your own way. And maybe reading books isn't your thing. There are plenty of other options. For example, many people like Moment in Time Learning from various digital sources such as LinkedIn, McKinsey and so forth. Many prefer audiobooks to print books. Others get their learning by listening to podcasts.

Different leaders have different learning styles. The key is learning in whatever form works best for you.

Bottom line: Learn every day and allow yourself to unlearn as things change rapidly.

Unlearning

The massive transformational changes that organizations need do not come from incremental changes. As a result, many approaches utilized in the past will have to be abandoned.

Sometimes, it is equally important to unlearn and just give up.

Implementing a system or ERP to replicate your existing process isn't the same as transforming with technology. Giving up an existing, obsolete process is one of the most common examples of how we unlearn in favour of learning and apply new technology.

For the longest time, organizations thought about the implementation of ERP with a heavy focus on their existing, as-is processes, trying to make the technology fit the current state or some version of the current state. We didn't want to unlearn our legacy processes. This stubbornness resulted in massive customization, delayed projects, and difficult-to-maintain systems. I even know of organizations that had to start a second, entirely new project to move into a vanilla version within just a few years of completing the *first* go-around of massive customization. I consider this a massive waste, the result of our inability or unwillingness to unlearn.

Organizations should start with the ERP's built-in processes instead of starting with their existing process. These days, you start with the "to be" state instead of the "as is" state, which you spend a lot of time defining, then selectively confirming with respect to the as-is state. We start with the "to be" in mind and use the system as a template, because the concepts of *system* and *process* are very tightly coupled in the new solutions.

Another example would be AI and related use cases. System implementation traditionally always started with a detailed requirement analysis, followed by planning, building, testing and rollout. In a process based on a traditional engineering approach, it took longer to leverage value for the solution. This is still relevant for many large technology projects, but for AI projects, this will often not work.

In AI use cases, you often don't know the requirement very clearly because the project is innovative in nature and heavily dependent on available data. As a result, you have to start with the cross-functional team, utilizing small, agile sprints of two weeks, and discover your requirement as you are building it. This approach challenges traditional management and engineering principles, since this is more agile and involves a high degree of experimentation.

Unlearning and then relearning how you make changes in your organization is one of the keys to becoming a digital leader.

Be Curious, Get Uncomfortable

Roy T. Bennett said, "It's only after you've stepped outside your comfort zone that you begin to change, grow, and transform."

Sign up for opportunities that make you uncomfortable! Almost nothing is impossible in this world. You just need to step out of your comfort zone and take up challenges.

Imagine if our ancestors hadn't been curious. They would never have discovered how to make fire by rubbing two sticks together. Curiosity is a mercurial quality that rises and falls throughout our lives, depending on what we are doing, where we are, and who we are with. This is both reassuring and daunting—reassuring because it turns out that curiosity helps us learn new things, and daunting because doing so involves a sustained and conscious effort on our end. **Every leader has to be consciously curious**.

Curiosity starts with the itch to explore. A 1964 study found that babies as young as two months old, when presented with different patterns, show a marked preference for the unfamiliar ones.[81] This instinct to explore grows into an instinct for inquiry. Sometime after their first birthday, children start to point at things, looking up at their parents as they do so. One of the main reasons babies point is to signal interest, to say, "I want to know about that. What is it?" Before they are able to speak, they are asking questions through rudimentary gestures. So, don't kill that inherent instinct to ask questions, experiment with new technologies, or explore new avenues.

In 2007, Professor Michelle Chouinard analyzed recordings of four children interacting with their respective caregivers for two hours at a time for a total of over 200 hours. She found that, on average, the children posed more than 100 questions every hour. Those children who were answered and had their curiosity met learned faster than those who did not. But, as Paul Harris of Harvard University points out, asking a question requires an impressively sophisticated mental process.[82]

Just like a child, you have to first realize that **there are many things you don't know**. There are invisible worlds of knowledge you have never visited. And you simply need to venture there, in that uncomfortable, unknown territory, like children do.

Make your mind curious. Try to answer the questions that result from your curiosity. Reach out to knowledgeable people for answers, read a book, or access some of the free, high-quality knowledge available on the internet. This is as much about learning as it is about training your brain and being as curious as a baby.

You must connect with the digital world with an original curiosity and constantly break down your known boundaries to enter unknown territory. It might be uncomfortable, but it's never unachievable! For example, using technology to do something like a monthly forecast or employee expenses saves you the time you would usually have to use to enter the data manually. It may take time to learn at first, but taking that time will quickly pay dividends and save you time in the long run.

You can be part of the change, or just have the change affect you. The whole point of becoming a digital leader is to be part of it—to be front and center within the change. Hence, you must familiarize yourself with and be comfortable practising "the art of the possible," because that's where the world is heading.

Initially, embracing new technology may feel over-whelming. I've heard it before: "Now we have to use some *app* and trust *technology* with less human over-sight and control." But if you are comfortable with trying it, the benefits are great. You get more time, which you can put toward something better for yourself and your organization.

Take Uncomfortable Assignments

Sometimes, getting outside your comfort zone will mean taking those uncomfortable assignments. As they say, "Sign up, learn, and make it happen." When those assignments come up that no one wants to touch, you've got to raise your hand and get uncomfortable.

Instead of thinking that you *have* to do this, change your perception. You *get* to learn. You *get* to raise your hand for something that will be a positive change, and you *get* to learn along the way and help make it happen. Isn't that the sort of opportunity you've been waiting for? Again, the whole point is to be front and center, within the change.

When I was with IBM several years back, there used to be something called the "stretch assignment." The idea was for curious high achievers to raise their hand and sign up for assignments above and beyond their day job/assignment. This discretionary effort was meant to be put forth on your own time—not the company's—and it was meant to stretch you. The key was that the organization encouraged it, and the accountability of balancing the time and discretionary effort was on the employees.

I really liked stretch assignments because I felt like I could raise my hand for anything, and I had more license to fail while still exploring and learning.

Organizations should consider adding "uncomfortable assignments" on top of the defined responsibilities.

↳ Get into uncharted territory to test new technology that you might not be comfortable handling.

↳ Remain curious, act consciously, take responsibility, and create experiences.

↳ Share experiences and accept others. Nothing is impossible. You can become whoever you want to be.

↳ We don't always know what we want to do, but by being curious and experiencing new tasks, you allow yourself to learn, understand better, and become a digital leader.

Don't be afraid to step out of your comfort zone. It only gets better if you venture into unknown territory. If not, how will you know what lies on the other side of the border?

Apply Technology in Everything You Do

If you don't maximize the use of available technology in your daily business work, someone else will go for it, and you will be left far behind. So, as a leader, what should you be doing?

Regardless of your department, **taking a technology-first approach is a basic requirement of your job and your personal life, too**.

Have you ever seen someone panicking because a laptop has been stolen, become lost, or crashed? It's not a laptop problem; the problem is that its user wasn't connected to the cloud. (These days, most digital immigrants have a Chromebook or something similar and everything they do is on the cloud by default, but it still happens.)

As an aside, a fellow programmer I used to work with had the amazing ability to lose almost anything. He did this regularly. He would often lose his car keys and would be searching on the internet for several minutes before he realized that he wasn't searching for a piece of code, and it wouldn't work. We used to have fun with it. But today, location tracking on objects is commonplace. I guess my coding buddy wasn't off-base.

The technology isn't the problem. It's how you use it. The same goes for your personal life. If you are a digital immigrant and have digital native kids at home, you have to apply technology to connect with them. A simple text goes a long way.

In the middle of the COVID-19 crisis, millions of people around the world within a very short time pivoted to using technology to work from home or remotely. For me, this confirms two things: 1) we have to apply technology in almost everything we do, and 2) when humans put their hearts and souls together and partner with technology, magical outcomes happen.

We will have a new normal on the other side of the crisis. As a digital leader, it will be important to be in front of it.

Work Smarter

Work equates to one-third of your life, and you cannot be an authentic leader if you act one way at work and are a completely different person at home. It is one continuous spectrum, so you must apply technology across the board.

⤷ Take a hard look at everything you do with minimal or no technology and consider whether you are compromising time, quality or money by doing so.

⤷ Pick up the phone and make a few calls. Ask how others are doing it. Most likely, by your third call, you will have a different perspective.

⤷ In our group, digital leaders have a habit of calling experts and other organizations to learn from them regularly. We learn from others and hypothesize for our needs, which is very powerful.

Are you only using email to connect with your stakeholders? Is that enough in the age of Instagram and Slack? Maybe not. Consider using these collaboration tools. Get uncomfortable and connect individually with your stakeholders to build an innovative team of digital leaders.

Are there certain activities that you or your team work on every day, month, quarter, or year, and *every time* you dread how painful and error-prone the process or activity is going to be? You are not alone. The question is whether you simply complained about it—perhaps vented about your IT department—or did something about it, such as talking with others in your organization to brainstorm some quick and easy solutions. Have you ever called someone from another organization to find out how they are doing it? Or have you Googled the problem to see what options are available?

Chances are, if you have a problem, someone else has had it, too. There are many solutions—large, medium and small. Robotic process automation (RPA) or data analytics-type solutions are often very fast and can address some of an organization's process and activity automation needs.

Now that most of our wealth and assets are electronic in nature, do you struggle and stress over the hassle of managing your million passwords? I did. Ask if your organization has an option for managing these. Maybe you yourself need to subscribe to one of the available solutions that will store, manage and even generate a strong password for you for your home and work needs. It sounds unreal, but I used to stress about this topic big-time! Fortunately, there are many solutions, such as Lastpass, Dashlane, RoboForm, 1password, etc.

Overall, technology should be a part of everything you do. Applying technology to business problems is a continuous process, and problems change rapidly with time.

⚷ Key Takeaways

- Unlearning is as important as learning. It is difficult to give up existing practices, but it is essential for reimagining your business.
- Be curious and get uncomfortable. If you are uncomfortable, you are learning for sure. Make sure you raise your hand for "uncomfortable assignments."
- You cannot be a technology-savvy person solely at work or solely at home; you need to embrace technology in all aspects of your life.

"Apply a technology-first mindset in everything you do, in and out of work. If you don't, someone else will do so—and you won't be there to witness it."

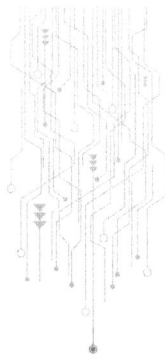

Chapter 8:
Human-Centric Thinking

While reading the book *How Great Leaders Think,* by Lee Bolman and Terrence E. Deal, I came across a wonderful line that emphasized the importance of putting your heart and soul back into the organization to which you are attached: "For an organization, group, or family, soul can also be viewed as a resolute sense of character, a deep confidence about who we are, what we care about, and what we deeply believe in."

Digital leaders don't just work with gadgets and AI; they also do so *with humans*. Being a human-centric thinker will help a leader keep her tribe intact through shared interests and communication. To be a human-centric leader, I would suggest, means thinking outside-in, **empathizing with stakeholders**, and always **focusing on the usability of everything you build** at work or at home.

Think Outside-In

The workplace of today is different from the workplace of the past, and the future will indeed be all the more different.

Tomorrow's workplace will be characterized by ubiquitous digitization, disorientation in the face of constantly changing technology, and the continuous erosion of stability and familiarity. Life and work alike will be fluid and episodic, with limited touchpoints between individuals and the organizations where they work.

Many people will feel a need for belonging and a desire to make a difference. This is where a human-centric digital leader steps in, **thinking from the perspective of others**, whether a client, supplier or colleague.

The process of putting yourself in their shoes and thinking as they would has amazing power and will help you analyze many factors. While leaders of the past kept their distance from their subordinates, both physically and psychologically, **leaders of the future need the involvement of everyone in the leadership process**.

Once upon a time, centralization was synonymous with control. But a leader today has to challenge the old format and nurture focus and coordination. Leaders must enable and support complex networks of relationships.

An **outside-in perspective** is required to counteract the bias toward inward focus. This is a deliberate activity, and it's an approach that aims to drag a business back into focusing on what is going on in the world of its stakeholders.

The acronym WIIFH (what's in it for her/him?) should be the focus, with "her/him" referring to your stakeholder. Digital leaders must be obsessed with WIIFH, as there is no better way to make your stakeholders successful. If your stakeholders are successful, it is most likely that the organization is turning into a successful TFO. Here are a few points to keep in mind moving forward:

- ↪ The details of how internal and external stakeholders make choices are difficult to isolate using data and analytics alone. Perspective and bias often play a larger role.
- ↪ Focus first on WIIFH for your stakeholders and on putting forth an effort to understand the problem from the stakeholder's perspective.
- ↪ Seek feedback at the moment.
- ↪ Listen to stakeholders' feedback at all key touchpoints.
- ↪ Find solutions and act on what you hear.

These are vital behaviours to practise, but even more important than practising them is to ensure that you are authentic about it. Once you learn the art of thinking from your stakeholder's perspective, you will connect better than ever before.

Setting the Stage

Adopting an outside-in perspective isn't as easy as it sounds. To make it easier, we treat it like the set-up for a play—with the customer as the protagonist.

Cast of Characters: The Customer and Other Actors

Like the hero of any play, customers are not alone. Their relationships with the other actors are central to how they feel, what they do, and the decisions they make. An outside-in perspective aims to understand these actors and the influence they have on the customer.

In business, the influencers will include suppliers, partners, and competitors. Consumers are influenced by anyone from family and friends to trusted intermediaries and government bodies.

Context: The Customers' Environment

Stories happen somewhere specific in place and time. In other words, they occur in a specific context. Understanding your customers' context is vital to understanding their actions and preferences.

An outside-in perspective deepens your understanding of the customers' context. Context will define business practices and consumer preferences or highlight anxieties (such as, for example, distrust of banks in Brazil due to past experiences).

Dilemma: Factors Affecting the Customer

Stories have drama, and a play or a movie always features an event that drives the plot forward. For most customers, there is some major factor in their lives or at work that is influencing their actions and concerns. For business customers, it may be new regulations; for consumers, it could be interest rates.

New technologies and economic trends affect customers. Understanding how these changes affect customers is essential to understanding their perspective.

Expectations: The Customer Experience

Knowing the genre of a play (adventure, romance, or comedy) does not necessarily mean we know exactly how the story will play out. Similarly, knowing the business (retail, finance, or utility), along with the characteristics of this sector, does not mean we know the specifics or the context of the customer story. An outside-in perspective uncovers how customers experience a business in relation to others.

Opening Night: Cast, Context, Dilemma and Expectations

Imagine pitching a concept for a play without knowing the key factors listed above—cast of characters, context, dilemma, audience expectations/experience and so on. It would be impossible! The same goes for businesses.

Knowing your customers, the people in their world, the dilemmas they face, their expectations and past experiences will enable you to tell a much better story about where you are heading and how you will take your customers with you. This approach will enable digital leaders to connect with stakeholders and increase the likelihood of success for their products and services manifold.

Do Not Assume—Empathize

I once had a really smart technical leader in my team. Let's call him Mr. Talented.

Mr. Talented was working very hard and had produced exceptional results. All of his projects and operational responsibilities were being delivered above expectations every single time, and his team loved him. My natural inclination was to recognize his performance and results through a promotion, which is often the obvious action a leader takes to recognize extraordinary performance.

I was about to start working on Mr. Talented's promotion, but I observed that, although he was performing exceptionally and was being recognized, he seemed increasingly stressed and unhappy. So, I decided to spend more time with him. I had two objectives:

- ↪ Make sure he was doing well on a personal level.
  - ■ I knew he had a young family with two children and a lovely wife. I didn't want to add pressure to his health and family life.
- ↪ Make sure that a promotion would be meaningful to him.

After a few conversations over the phone, I sat down with him in person to understand what was going on. It turned out that the reason he was so worried and stressed was that he'd anticipated that I would offer him a promotion and a new role to lead the bigger team, which would mean a lot more travel and administrative responsibilities. He felt he wouldn't have a choice in this matter. It soon became clear that he really didn't want to grow in terms of hierarchy. He loved people and worked well with them, but he also loved technology. He actually preferred to work with technology, including its more challenging aspects.

I decided not to follow through on a promotion path for Mr. T. Rather, I worked on putting together a one-time bonus and allowed him to continue working in the same job, which was going quite well for him and the organization as a whole.

After our conversation, I saw a much happier Mr. T. I had been on the verge of stereotyping him as an individual who was seeking advancement through promotion—instead of seeing things from his perspective, which told a much different story. Mr. T. has since evolved into a more mature professional and has successfully led several innovative initiatives.

There are two lessons to draw from this story:

1. Leaders should take the time to listen, understand and empathize with stakeholders without falling into the trap of stereotyping.

2. A digital leader like Mr. T can lead larger innovations *without* managing a larger team. Growth isn't necessarily proportionate to the size of the team and title.

Empathy is important for any leader when it comes to navigating the human elements of leadership and leading with humanity. Hence, **assuming without asking yourself why someone did something** is a no-no for a human-centric leader.

Leaders may occasionally find themselves baffled as to why a member of their team made a certain decision, is behaving a certain way, or neglected to act appropriately in a situation. Self-awareness of our own bias is critical, and there's no harm in asking that extra question or doing a bit of analysis where it matters. This starts with getting to know yourself better. Then you can become more conscious of how your emotions influence your behaviour—and how your behaviour influences other people's emotions. This includes the ability to identify what shapes your opinions and avoid projecting them onto other people or even jumping to assumptions without understanding why someone did something you don't approve of.

It can be difficult to understand the way others think and feel, but working on this will help you to best address the needs of your team members in an effective way. Let's say, for example, that one of your employees is suddenly demonstrating disengagement, and their frequent absenteeism is causing undue pressure on other members. Leaders always have a choice in how they react to such situations, and the route they take often defines them.

The world really isn't that bad. Most people come to work to be useful and to contribute meaningfully. Leaders of organizations can make it harder or easier to do so.

The Dangers of Bias

In Chapter 5, I mentioned the issues that can arise when a job candidate has material that is inconsistent with their social media presence and their resume. The other side of the coin is also interesting. Our bias can stereotype candidates based on social media. We have to be careful.

I'm usually a quick study when it comes to interviewing. I usually do my research before the interview and usually get a good read on the candidate within the first fifteen minutes. But I acknowledge that I have my bias, too.

I am agile, and I tend to value speed. Once, I was interviewing an architect in what was supposed to be the third and final round of interviews. Usually, this round is a "culture check" kind of interview, meaning the deeper technical and HR interviews would have happened by then.

Although my team strongly recommended this candidate, I wasn't feeling confident, based on the interview. I couldn't help but notice that the candidate was pausing a lot before answering my questions. By the end of the interview, I was almost positive that he was not the type of hardworking, fast professional we were looking for, and I was quite hesitant to move forward with this hire.

I make a lot of decisions based on my gut feeling, and in this case, my gut was telling me not to proceed with the hiring process. But knowing when to slow down and listen to others is also important. The hiring manager was extremely confident about this candidate. Based on the confidence of the hiring manager, who worked for me, and all the other folks who had participated in the other interviews, I decided to support the decision to hire in spite of my misgivings.

This time, it turned out that my gut feeling was wrong. This architect proved to be one of our best hires. If I had allowed my own bias to overrule the informed opinions of my team members, our organization would have missed out on a strong addition to our team. I'm glad I went with the objective approach in this case vs. going with my gut.

The point is, we all have biases, and it's important to be aware of that.

Focus on the Facts

Paying attention to the facts is an important attribute for a digital leader. A clear mandate for leaders is to restructure the insights being gathered to include qualitative, human-centric elements.

It is traditional to conduct employee surveys every couple of years to understand the employee mindset and engagement level. But remember: this is based on questions that are created top-down.

Instead of conducting your survey on an arbitrary date (e.g., every two years), consider collecting insights at the moment when it really matters. These days, this can be achieved quite effectively via technology. Timely insights may prove much more valuable in our fast-changing digital world.

It is important to understand the context to make each insight useful and actionable. Digital leaders should take an active interest in understanding context.

Modern challenges call for ongoing information-gathering. Moments matter. Continuously gathering insights is no longer a "nice to have" feature; it's a necessity.

There are multiple ways to gather information and insights from your team. I advocate for doing it ethically, with the intent of utilizing a WIIFH mindset. Include human-centric questions that focus on your teammates and clients as individuals.

By using a human-centred approach as the core of your leadership philosophy, you are developing yourself as a leader who is **accessible**, who **celebrates diversity** in people and thoughts, **welcomes**

opposing viewpoints, is **flexible**, is **open to change**, and who **respects and values** every person you interact with.

Leaders who engage in this type of thinking will be able to shape the emerging future.

Focus on Usability

Usability is everything, and it's more important for a product or service to be intuitive than pretty. This is true of all the products or services you build or support. That's why you hear so much about "design thinking" and "user experience."

My mom is a digital visitor. She never used a computer in her life until tablets came onto the market. Now she spends more time on her smartphone on Facebook, WhatsApp, video chat, etc., than the rest of us in the family combined. She gets grumpy if we talk about some activity her grandson has done today and she hasn't already received a Facebook or WhatsApp update or a picture of it. It's almost like we need to send her materials to read before any discussion—like meetings in the workplace.

It's funny to think that this is the same person who, about fifteen years back, when I moved to North America, couldn't get on Skype for a simple video chat. The steps of switching on the computer, making sure the internet connection was working, and then clicking on the Skype icon were too much for her.

So, what changed?

For a digital visitor, there is a massive difference between a laptop and a tablet. For my mother, using a tablet means picking up a light device in one hand—a device that is always on—logging in with her thumb, and placing her finger on the Facebook Messenger icon. This is quite a bit different than the experience of using a laptop. The success of mobile devices isn't just due to convenience but also to their intuitive usability.

Consequences of Subpar Usability

I'm not convinced that every single-purpose device will eventually be replaced by a single convergent device, but I'm pretty sure that no one will ever make another single-purpose device as bad as the TwitterPeek.

Released in 2009 for a whopping $200 price tag, the TwitterPeek was designed specifically to let you check your Twitter feed—that's it. The designers also decided that you only needed to see twenty characters of a tweet to know if it was worth your time ... so each tweet needed to be opened individually to be read. The only other thing you could do with the device was to send more tweets. This was never, ever going to be an iPhone killer.

Google X (2005) is another example of poor usability. Time for online interfaces! Google introduced, for just one day, a totally crazy Apple OS X-style dock above their search bar. It featured ugly icons, was an architectural mess, and lacked a clear connection to any use case. The interface was dropped less than 24 hours after its introduction.[83]

Google Wave was a web-based computing platform launched in 2009, announced as a failure in 2010, and killed in 2012. Wave was an interesting project that was intended to improve collaboration, and I remember how excited I was to get access to it. Unfortunately, Google forgot about user experience, and it turned out to be a design nightmare. What a pity.

We often encounter websites that look pretty, but if we can't find what we want, what's the point? Useful is more important than pretty. We like apps because an app does just one or two things. You go to an app on your phone for a particular purpose, to achieve a specific result.

The explicit promise is that a product or service will help us achieve a certain result; the implicit promise is that it will be easy and usable.

Usability takes care of the implicit requirement and makes it useful for the users, in addition to providing useful features as per explicit requirements.

Even common household products are subject to the rules of usability. When I was growing up, we used to have ketchup bottles made of glass with a cap on the top. Often, you couldn't access the ketchup at the bottom of the bottle. The solution was to leave the bottle standing upside-down so that the contents at the bottom would slowly work their way down. As a result, bottles often broke due to accidental falls. Today, many household items—not just ketchup but toothpaste, shampoo, etc.— come with wider, flatter caps that can act as the base of the bottle for it to stand. The flatter cap encourages us to stand the product upside-down, meaning gravity is working for us rather than against us, always pulling the product toward the cap and making usage easy.

This may sound like a silly, simple example, but at its core, the main driver of this change was to provide a better user experience. These organizations started using customer needs/pain points as the building blocks of their product design.

Functional over Features

In 2000, I joined an interesting project focused on building a software solution for a very well-known and accomplished engineering guru from Japan (let's call him Dr. M.). Dr. M. was a sought-after consultant for major corporations like GM, GE, etc., and was one of the brightest people I had ever met.

I was selected for the project due to my math and physics background, and we spent ten days in a boardroom. With 10 to 12 hours of math and statistics every day, it felt like a mini master's program. A few of my colleagues and I were very excited. Interestingly, this engineer knew what he was doing and could do all of it in Excel—anywhere, anytime. His exposure to other types of software was limited, though, and that was where my team would help him build a solution.

Our team of IT professionals understood the software, and we were only too eager and ready to jump in and code. But in our eagerness, we didn't stop and talk to the engineer about what his daily life was like and how he would be using the software We should have considered what it would be like to be a fly on the wall watching him deal with his clients. But we didn't. And that was a big mistake.

Without taking the time to empathize and understand the context, we built a very robust web application.

After some rigorous testing, it was doing complicated computer-aided parameter design (CAPD), quality loss functions (QLF), and design of experiments (DoE) really well and coming up with excellent recommendations. Unfortunately, when we started rolling it out to the engineer, we discovered that we had to do a lot of reworking to make it acceptable to him.

What happened? Our software had all the core functionalities the engineer needed. Why didn't it meet the customer's needs?

The issue was not the core functionality. We had built a web application at a time when wi-fi wasn't as broadly available as it is today. A lot of the work was done on shop floors. Dr. M would often do a lot of work on planes, but, as our software required an internet connection, one couldn't use it on a plane in those days. He liked to carry a light laptop with him as he travelled, but a typical laptop back then didn't have a lot of horsepower. He needed to show or share his analysis results with clients without them needing to have access to the software, but one could only share or open the results when logged into the software over the internet.

In other words, we had fully functioning software that was not useful to the user.

Once we understood the context of how Dr. M. was using the software, my team started reworking it. We switched to a desktop-based light application that used binary files for storing the results and Excel-based outputs that could be consumed by the clients.

In those days, we used to call it "nonfunctional requirements." Today, to put more emphasis on the user, we call them "user experience requirements." Usability is closely connected with WIIFH thinking. It's about how and why it will be used.

Any time you are thinking about developing services or products, you must think about the people you are in contact with. How do you interact with them? Are they at the core of your decision-making? After all, they are your users. Look through their lenses, not just yours.

KeyTakeaways

- Think "outside-in" in everyday life. You will connect with your stakeholders more than ever before.
- Do not assume. There is huge value in putting some extra effort into asking and listening.
- Focus on the *usability* of everything you produce as a product or service—whatever you provide will be more successful.

"If you want to win the hearts and minds of others, be part of their bias."

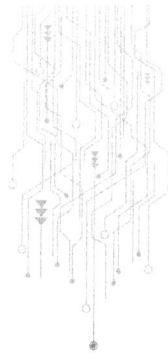

Chapter 9:
Agile Leadership

The title of this chapter may not be immediately clear, as there are many interpretations of the word "agile." It often refers to the methodology of scrums and sprints. The context here is all about being flexible and willing to adopt changes in the ecosystem.

It is essential for digital leaders to be agile in everything they do. The environment is changing faster than ever, and hence, digital leaders need to be agile to be productive. The value of an early start is greater than ever before.

I would like to start off by quoting Filippo Passerini, one of the leading global digital leaders, P&G's digital czar and president of Global Business Solutions. He once said, "We intentionally put the cart before the horse, because it is the way to force change."

A key aspect of competing in the Digital Age is the ability of leaders to be comfortable with a certain level of ambiguity when it comes to digital initiatives. Leaders should be willing to be trailblazers even if they don't fully understand the technology yet—putting the proverbial cart before the horse.

Build as You Define

Normal engineering will teach you that the correct process is to define, then engineer, then build. This discipline of following a proper sequence is important and works for many things in life.

But the traditional, sequential way of doing things doesn't always work. Often, we have to create a cross-functional pod in which you are actively building while you are defining. That's the thing; you have to be comfortable with a different kind of engineering that isn't always sequential.

Over the course of our journey through Industrial Revolutions 1, 2 and 3, we perfected mass production, and we matured in efficiently delivering consistent services and products through strong engineering and continuous improvement. It was about producing a defined, consistent business outcome.

For decades, we really looked up to Japan for its culture of continuous improvement, featuring tight processes with minimum deviation. We often hear that Japanese cars are of higher quality and hold their value. Most of that is attributed to robust engineering and quality control. Making a car involves a highly sophisticated assembly line. This presents well-defined problems that were solved very well following a traditional assembly line system. But now, since the reimagining factor comes in, that doesn't always work.

Now, we are grappling with undefined or not-so-clearly-defined problems and opportunities. That's why I think the basic philosophy of engineering will also change. It has to change to a more agile kind of model. That's where a leader needs to be an agile driver, so as to set the approach in the correct direction and constantly adjust in conjunction with business needs. From identifying transformational improvements to finding combinations of process and technology to make that happen, a digital leader needs to have the flexibility to adjust the usual prototypes and strike it big. Remember: **Think big; act now; start small; scale fast.**

Implementing Agile Leadership

Today's volatile and unpredictable business landscape, shaped by digital disruption, requires leaders to be **agile drivers of innovation**. Such leaders need to exhibit high levels of humility, adaptability, vision and engagement through hyper-awareness, informed decision-making and quick execution. Combining these must-have competencies will ensure that you are well equipped to deal with today's disruptive business environment.

Implementing agile leadership involves **turning the company pyramid upside-down**, allowing those closest to the customer to make better decisions based on customer needs. If you wish to be an agile driver of change, you need to do it *now* and not wait for tomorrow.

Stick to the 3C's—**communication**, **commitment** and **collaboration**. Your sleeves should always be rolled up, and you should always be prepared to deliver the needs as quickly as possible.

Planning for the future and delegating work are signs of a good leader, but a leader also isn't afraid to take immediate action. This can mean doing something now—even before it's not completely understandable or clear to people, even if it means starting without complete planning or the ability to launch a prototype or a pilot—with the "big picture," as they call it (meaning the vision).

This is critical for leaders for multiple reasons:

1. **Time**. Getting to market on time is essential because almost everything can be replicated. That means the earlier you go to the market, the greater your advantage to capture the audience. (And, by the way, that's what Amazon did for years. Amazon focused on revenue, not on profitability, because they wanted to capture the audience.)

2. **Urgency**. Start now, and do what needs to be done, even if you need to learn the skills as you go. If you're doing a plumbing project, be the plumber for the day.

Tomorrow Equals Today, and Today Equals Now

Now is more powerful than ever before. Agile, hyper-aware leaders are focused on spotting emerging digital opportunities or competitive threats immediately. As Alexander Dahm, the vice president of Space

Equipment Operations for Airbus, says, "There needs to be an everlasting radar, a high-speed radar, where you constantly track the situation and act immediately, not waiting for tomorrow to happen."

If you're an agile driver of change, you need to be engaged, seek new insights, and adapt in response to the market. And as you adapt, you must be aware of the need to provide guidance to your team, your employees, your customers, and even the backend suppliers.

You need a strong vision to achieve this, as the potential for change threatens to overwhelm a linear strategy. When your strategy is dependent on market conditions and external factors—all of which are beyond your control—agility allows you to shift and move based on those factors.

Take, for example, the 2017 Equifax breach, during which hackers were able to steal personal credit records from millions of people.[84] Equifax chose to wait for a month before disclosing the breach. This left the door open for wrongdoers to take advantage of the time lapse between the company's discovery of the breach and the public's knowledge of it. The press and Equifax's competitors were given a distinct advantage.[85] Equifax didn't realize how critical every hour truly was, and it cost them.

On September 26, 2017, three weeks after Equifax disclosed the data breach, CEO Richard F. Smith announced his retirement, explaining in a statement that, "At this critical juncture, I believe it is in the best interests of the company to have new leadership to move the company forward." In addition, Smith did not receive his annual bonus.[86]

If you had a breach in a brick-and-mortar building, you could secure it. You could block the doors and windows and close off the roads to control the damage. In the case of a cyber breach, as long as the opening is there, there are millions of roads in and out of the "building." If you really want to stop it, you are putting a halt to your business, too.

Every moment counts, and good leaders are constantly scanning their environments, inside and outside their organizational boundaries. With technology-driven change accelerating across industries, it has never been more important for leaders to take immediate action.

Keep in mind:

↳ Fast execution entails willingness on the part of a leader to move quickly, often valuing **speed over perfection**.

↳ In an environment characterized by significant disruption, **the effectiveness of informed decision-making is significantly reduced if the leader isn't able to act quickly**.

↳ Ultimately, agile leaders will only be effective if they are able to **execute quickly, based on partially informed decisions**.

↳ There are many barriers to agility in business, be they organizational, fiscal, structural, or cultural. At times, you might even have to shorten bureaucracy, as **fast decision-making requires fewer signatures**. That's adaptability in action.

Think Big, Act Now, Start Small, Scale Fast

Business in the Digital Age may seem complicated, but I have always known that the difference between success and failure lies in eight simple words: **think big, act now, start small, scale fast**. If you follow this rule, even the stoppable becomes unstoppable.

By thinking big, successful innovators consider the full range of possible futures. Make sure you understand emerging technology in the proper context, rather than within the bounds of current assumptions. Rather than simply looking for faster, better, or cheaper products, you need to dare to dream big. Successful innovators are those who are willing to start from a clean slate, to pursue new products that could rewrite the rules of a niche or even an entire industry.

Take, for example, the self-driving car team at Google. They're not just making better cars but focusing on full automation, trying to take human drivers out of the loop entirely, as that will dramatically improve usage patterns and disrupt business models. By contrast, those who fail to seize the big opportunity typically think small. They assume that the future will be a slightly different version of the present.

Successful innovators **start small after thinking big**. Rather than jumping on the bandwagon for one potentially big product, they break the idea down into smaller pieces for testing. Leaders who don't start small tend to panic in the face of disruption.

Agile drivers of change will also make their organizations scale fast by displaying the attitude that **a demo is worth more than thousands of pages of business plans**. You have to conduct extensive, inexpensive prototyping before you even get to the pilot phase—let alone the big rollout—so you can gather comprehensive information and quickly analyze both what's working and what isn't.

Never forget:

↳ There is no need to assume that customers will stick with you forever. Microsoft, Motorola, Blackberry and Nokia all missed the smartphone business because their technology assumptions didn't fit with new changes. They couldn't envision how such changes might challenge their own products.

↳ Defer important decisions until you have real data. Google's early investments in its driverless car weren't all that much more than what car companies typically spent on Super Bowl ads. Google started with relatively small investments in the beginning, then worked its way up to a larger spend.

↳ It is great to be passionate about your project. But you must have the discipline to set aside or alter projects based on what you *learn*, not on what you or others *hope*. Ask the hard questions. Fail fast if you must. This way, you can course-correct and proceed forward.

↳ Thinking small, starting big, and not scaling quickly is what killed Blockbuster. It ignored Netflix's DVDs-by-mail model for years, then bet big on its own version before fully working out the economic and operational implications. It turned out that Blockbuster's business model couldn't handle the loss of those hated late fees.

↪ Deliver tangible, smaller outcomes frequently, but keep the end in mind. A tangible outcome is integral to learning and honing digital skills that also have an impact on the daily roles of employees and what they can provide to the organization in the long run.

Roll Up Your Sleeves

Traditionally, your job description determined your role in any given project. We said things like, "This is my role in your project, and this is your role." But if you have the fire in you to learn, it is absolutely necessary for you to participate in anything and everything and to contribute wherever you can add value.

Digital leaders aren't rigid about their job descriptions and traditional roles. Instead, they're willing to jump into new opportunities and innovate, even if it's not always part of what their job title entails. They create the non-hierarchical environment to foster innovation. I always propose that digital leaders must behave like scientists; they are out to innovate. Digital leaders innovate and, thus, find new strategies for the future. And just as a scientist has his nose in everything, seeking the unknown, you too need to be engaged in territory where you are "not supposed to be."

Nothing is below your pay grade. This digital world is based on knowledge. Titles and hierarchies aren't as important as they once were. The point is, wherever and whenever you can add value, even if it's not in your job description, everyone will benefit.

Think about the startup. When a company is in startup mode, there are one or two founders and a few other key hires to think about. Picture a small group of people in a room, chasing an idea and a customer obsession, and do that. Amazon calls this approach the two-pizza team. For a new idea, they create a team small enough that two pizzas can feed them. No more and no less than that.

The kind of mindset we're talking about is a two-pizza deal. And whoever is in the two-pizza team, regardless of their title, they're chasing a product or an idea.

When you're choosing a product and an idea in a cross-functional way, in a lateral way, everybody is a doer. That means you're always ready to do and contribute as much as you can, regardless of your predefined job title. **Everybody, regardless of their traditional roles or skills, has one job: make the purpose successful**. You show up ready to contribute appropriately and extensively to whatever you can, with a fire in your belly and a hunger for the unknown. Always be ready and willing to take the plunge.

Agile leaders:

- ↪ **Recognize that leadership is to be found everywhere within the organization**.
  - ◼ But leaders also need to provide the opportunity for people to lead themselves as well as

others. This enables the agile culture to spread further throughout the organization. By setting an example and showing everyone what agile leadership looks like, you can sustain a cultural change that will eventually become the new norm.

↳ **Empower individuals and teams so they can decide and act upon things themselves, within their own boundaries**.

- To this end, agile leaders need to trust their teams and dare to let go of power and control. Remember that you need to turn the pyramid upside-down! The teams are not there to serve their leader; the leader is there to remove obstacles that prevent the team from performing at its best.

↳ **Realize that the people closest to the customers and the customers' problems may have the best insights into possible solutions.**

- An agile organization ensures that there is a good flow of ideas and that actions are taken on these ideas. While this doesn't mean that all ideas will be used in practice, it inspires a culture of continuous innovation.

Key Takeaways

- Tomorrow = today; today = now. *Now* is more powerful than ever before. Time to market is one of the most critical aspects of digital leadership.
- Think big; act now; start small; scale fast. Always be reimagining your organization and your place in the market.
- Roll up your sleeves and get engaged. Always be willing to learn and get your hands dirty.

"Speed is money."

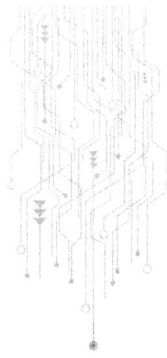

Epilogue

Digital is no longer the responsibility of a few tech-savvy executives. Rather, in this Digital Age, every person involved in a company's daily execution needs to be part of the digital revolution.

Leadership 4.0 asks of you to **invest 50 deliberate hours over 100 days in yourself**, regardless of whether you are optimistic/pessimistic or whether you're a DN, DI or DV. Once you invest those hours, you will surely rediscover yourself as a "digital leader," and the exercise will contribute more to your professional development than some executive MBAs.

Why not try out the plan below?

Step 1: Conduct a Self-Assessment

SELF-ASSESSMENT

1: Never	2: Rarely	3: Sometimes	4: Mostly	5: Always

Habit / Behaviour	Questions	1	2	3	4	5
Be Authentic. Be Yourself	How frequently do you feel comfortable showing up and acting as yourself, regardless of the forum and stakeholder you are with?	1	2	3	4	5
Be Authentic. Connect Individually	How often do you make the effort to get to know the people you meet with an intention to 'give'?	1	2	3	4	5
Be Authentic. Work the Talk	Would you say that your actions align with your talk?	1	2	3	4	5
Embrace Technology. Unlearn and Learn Technology	How often do you experiment with new technologies/apps and review/change /give up older practice, technology and habits?	1	2	3	4	5
Embrace Technology. Be Curious, Get Uncomfortable	In the past year, how often have you signed up for assignments which are beyond your past experience?	1	2	3	4	5
Embrace Technology. Apply Technology in Everything You Do	Thinking about your projects in the past year, how often have you applied technology instead of a paper/manual process?	1	2	3	4	5
Be Human Centric. Think Outside in	How often do you think: What's in it for him/her? ('WIIFH') before 'What's in it for me'(WIIFM)?	1	2	3	4	5
Be Human Centric. Do Not Assume	Do you always ask enough questions and demonstrate curiosity when you're unsure about available information?	1	2	3	4	5
Be Human Centric. Use the Lens of the User	Do you consider and put deliverable effort on improving usability of all services of the products you deliver?	1	2	3	4	5
Be Agile. 'Tomorrow = Today; Today = Now'	Do you strive to execute work in the earliest possible timeline?	1	2	3	4	5
Be Agile. Think Big; Act Now; Start Small; Scale Fast	Do you focus on the big picture? Are you able to get started quickly with a minimum useful product (MUP) and scale?	1	2	3	4	5
Be Agile. Roll Up Your Sleeves	Do you get involved with the team at the working level?	1	2	3	4	5
Total						

Step 2: Get Feedback

Identify a trusted mentor/colleague for a peer assessment of the above template.

Step 3: Evaluate and Adjust Your Approach

If your score is 4 or above for each behaviour, continue the good work and focus on the areas you want to further improve.

If it is less than 4 for some or all of the habits/behaviours, follow these recommendations:

Every day for 100 days, invest 30 minutes into the habit/behaviour, as follows.

First thing in the morning, take 15 minutes to:

↪ Review the habits/behaviours. Start with the ones with an initial assessment score of less than 4.

↪ Identify the behaviours you will focus on during the day and have an action plan integrated as part of your normal day.

At the end of the day, take another 15 minutes to:

↪ Review the behaviours you focused on for the day.

↪ Reflect and journal how you have worked on those behaviours during the day in a Journey Log.

It is important to continue the above routine over weekends and holidays. Remember that actions could be work- or non–work-related, as work and life are more integrated than you think.

I strongly encourage you to have a partner so that you can keep each other on track or even share your respective Journey Logs.

Once you have at least three journal entries of applied actions for each habit/behaviour, mark it with the sentence, "I am ready." Then aim to get all 12 behaviours in the "ready" state. Note that those that started at a 4 will require less time, and you should be able to journal actions fairly quickly. Others, however, might take multiple attempts. Don't forget that **change always starts with I**, and it's never easy.

Below is an example of a Journey Log template:

JOURNEY LOG TEMPLATE

Habit / behaviours	Applied – Journal – Habit Behaviour in Action	I am ready(Y/N)
Be Authentic. Be Yourself	‹Date›	
Be Authentic. Connect Individually		
Be Authentic. Work the Talk		
Embrace Technology. Unlearn and Learn Technology		
Embrace Technology. Be Curious, Get Uncomfortable		
Embrace Technology. Apply Technology in Everything You Do		
Be Human Centric. Think Outside in		
Be Human Centric. Do Not Assume		
Be Human Centric. Use the Lens of the User		
Be Agile. 'Tomorrow = Today; Today = Now'		
Be Agile. Think Big; Act Now; Start Small; Scale Fast		
Be Agile. Roll Up Your Sleeves		

Acronyms

Acronym	Stands for	First used on page
AI	Artificial intelligence	11
DI	Digital immigrant	29
DN	Digital native	29
DNA	Fundamental characteristics of an organization	31
DV	Digital visitor	30
ERP	Enterprise resource planning	99
ExO	Exponential organization	68
Fintech	Financial technology	52
FIS	Furthest imaginable state	86
FOMO	Fear of missing out	12
GDP	Gross domestic product	61
GWP	Gross world product	61
MTP	Massive transformative purpose	81
P-type	P-type (product): "a technology that was widely dismissed before ultimately triumphing"	22
S-type	S-type (strategy): "a new way of doing business, or a new application of an existing product, which involves no new technology"	22
SAP	Name of a software company	92
TDNM	Technology does not matter	30
TFO	Technology-first organization	26
TS	Technology scratcher	30
US	United States	24
USD	United States dollars	61

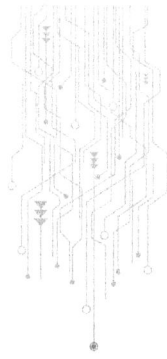

Acknowledgements

I would like to thank all my teachers, who helped me inculcate the learning of my life. By "teachers," I also mean my past and present colleagues, team members, customers, partners and authors of hundreds of books that I have read over the years. I learned from each one of you through these years, and I am indebted to you for the shared knowledge. A big thank-you to all of you for contributing to my thoughts and inspiring me to think better and to think differently.

I would also like to thank my wife, Jhumur, the first person with whom I shared the idea of writing this book. She just said, "Do it." Jhumur is one of the most positive people I have ever met. She supports every new idea of mine, no matter how crazy it might sound. Thank you!

I would also like to thank my friend Saheli, who guided me through this project and helped me edit my thoughts and expressions.

Finally, my biggest inspiration is my son, Rishi. We both love books, and we spend a lot of time reading. His inherent curiosity helped me throughout my research, inspiring me to start asking questions of myself. Those answers went into the making of this book. I love you more than anything else.

Being an avid reader myself, I would also like to mention the names of some world-renowned leadership books that have provided me with loads of examples and case studies, many of which I have referenced in my book. Books like *Exponential Organizations* by Salim Ismail, *Homo Deus*, the million-copy bestseller by Yuval Noah Harari, *Factfulness* by Hans Rosling, and last but not least, *Leading Digital,* by three well-known Harvard Business School academicians: George Westerman, Didier Bonner and Andrew McAfee. I am also indebted to several digital leaders whom I come across regularly and whose lectures have taught me a lot about the subjects I have discussed in this book.

Finally, I would like to thank the frontline professionals in healthcare, supply chain and technology who are relentlessly fighting the COVID-19 pandemic to sustain the planet, people and the economy.

About the Author

Debasis Bhaumik, more often known as DB, is an entrepreneurial leader who has worked with and for many global Fortune 100 organizations across Asia-Pacific, Europe and North America. He started his career as a technology entrepreneur in India and moved up through the ranks in North America. He is a well-respected senior leader in the digital transformation space.

DB's approach to life and work is consistent and based on "Think Big, Act Now, Start Small, and Scale Fast." He loves to call himself a lifelong student who reads at least one book every week from various sources. DB finds pleasure in continuous learning and exposure to the unknown but finds more pleasure in "giving" through mentoring, teaching and coaching. For over a decade, he has been a visiting faculty member at various universities in India and Canada.

DB loves his work. In his words, he "gets to have fun at work and also get paid." He believes that unlimited potential is out there for all willing and curious professionals in Industrial Revolution 4.0.

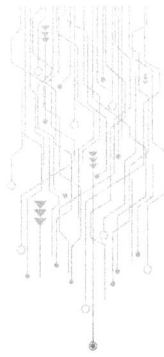

Sources

1 Schwab, Klaus. "The Fourth Industrial Revolution:
   What It Means and How to Respond." Foreign Affairs,
   December 12, 2015. https://www.foreignaffairs.com/
   articles/2015-12-12/fourth-industrial-revolution

2 Elsner, Ann. "Polaroid Inventor Edwin Land Gave Us More
   Than Just Instant Photos." Smithsonian Magazine, May 18,
   2018. https://www.smithsonianmag.com/innovation/
   polaroid-inventor-edwin-land-gave-us-more-than-just-
   instant-photos-180969119/

3 "Pan Am Milestones." The Pan Am Historical Foundation.
   https://www.panam.org/history-resources-museum-
   links/51-pan-am-s-aviation-milestones

4 "5 Companies With the Most Remarkable Digital
   Transformation Strategies." Zigurat. https://
   www.e-zigurat.com/innovation-school/blog/
   companies-digital-transformation-strategies/

5 Johnson, Khari. "Domino's Pizza bot now offers its full menu
   and custom orders on Facebook Messenger." VentureBeat,
   February 1, 2017. https://venturebeat.com/2017/02/01/
   dominos-pizza-bot-now-offers-its-full-menu-and-custom-
   orders-on-facebook-messenger/

6   Bahcall, Safi. Loonshots: How to Nurture the Crazy Ideas That Win Wars, Cure Diseases, and Transform Industries, St. Martin's Press: New York, March 2019: 116.

7   Clark, Travis. "Netflix is still growing wildly, but its market share has fallen to an estimated 19% as new competitors emerge." Business Insider, January 24, 2020. https://www.businessinsider.com/netflix-market-share-of-global-streaming-subscribers-dropping-ampere-2020-1

8   Bahcall. Loonshots, 10

9   Rosling, Hans and Rönnlund, Anna Rosling. Factfulness: Ten Reasons We're Wrong About the World—And Why Things Are Better Than You Think, Flatiron Books: New York, 2018.

10   TED Talks Library: Hans Rosling. https://www.ted.com/speakers/hans_rosling.

11   Musk, Elon. "Making Humans A Multi-Planetary Species." New Space, June 2017, Volume: 5 Issue 2: 46–61. http://doi.org/10.1089/space.2017.29009.emu

12   Shekhtman, Lonnie. "NASA Takes a Cue From Silicon Valley to Hatch Artificial Intelligence Technologies." NASA's Goddard Space Flight Center, Greenbelt, Maryland, November 19, 2019. https://www.nasa.gov/feature/goddard/2019/nasa-takes-a-cue-from-silicon-valley-to-hatch-artificial-intelligence-technologies

13   Kurzweil, Ray. "Sooner Than You Think, We Will Connect To The Cloud Directly From Our Brains." Fast Company, March 21, 2014. https://www.fastcompany.com/3027984/sooner-than-you-think-we-will-connect-to-the-cloud-directly-from-our-brains/

14   Garreau, Joel. Radical Evolution: The Promise and Peril of Enhancing Our Minds, Our Bodies—And What It Means to Be Human, Doubleday: New York, 2005.

15   Marquet, David L. Turn the Ship Around!: A True Story of Turning Followers into Leaders, Penguin Group: New York, 2012.

16 Frey, Carl Benedikt and Osborne, Michael A. "The Future of Employment: How Susceptible Are Jobs to Computerisation?" Oxford Martin School. **https://www.oxfordmartin.ox.ac.uk/downloads/academic/The_Future_of_Employment.pdf**

17 Harari, Yuval Noah. "The rise of the useless class." Ideas. TED.com, February 24, 2017. **https://ideas.ted.com/the-rise-of-the-useless-class/**

18 Shell, Ellen Ruppel. "AI and Automation Will Replace Most Human Workers Because They Don't Have to Be Perfect—Just Better Than You." Newsweek, November 20, 2018. **https://www.newsweek.com/2018/11/30/ai-and-automation-will-replace-most-human-workers-because-they-dont-have-be-1225552.html**

19 Muro, Mark and Maxim, Robert and Whiton, Jacob with contributions from Hathaway, Ian. "Automation and Artificial Intelligence: How machines are affecting people and places." Metropolitan Policy Program at Brookings Institution, January 2019. **https://www.brookings.edu/wp-content/uploads/2019/01/2019.01_BrookingsMetro_Automation-AI_Report_Muro-Maxim-Whiton-FINAL-version.pdf**

20 Harari, Yuval Noah. Homo Deus: A Brief History of Tomorrow. 1st Edition, Harper: New York, 2017.

21 Batista, Michael A and Gaglani, Shiv M. "The Future of Smartphones in Health Care." AMA Journal of Ethics, 2013; 15(11): 947-950. **https://journalofethics.ama-assn.org/article/future-smartphones-health-care/2013-11**

22 Naudé, Wim. "Artificial Intelligence against COVID-19: An Early Review." Towards Data Science, April 1, 2020. **https://towardsdatascience.com/artificial-intelligence-against-covid-19-an-early-review-92a8360edaba**

23 Bullock, Joseph; Luccioni, Alexandra; Pham, Katherine Hoffman; Lam, Cynthia Sin Nga; Luengo-Oroz, Miguel. "Mapping the Landscape of Artificial Intelligence Applications against COVID-19." Cornell University Archive. **https://arxiv.org/pdf/2003.11336.pdf**

24   Hao, Karen. "This is how the CDC is trying to
     forecast coronavirus's spread." MIT Technology
     Review, March 13, 2020. **https://www.
     technologyreview.com/2020/03/13/905313/
     cdc-cmu-forecasts-coronavirus-spread/**

25   Chun, Andy. "In a time of coronavirus, China's
     investment in AI is paying off in a big way." South
     China Morning Post, March 18, 2020. **https://www.
     scmp.com/comment/opinion/article/3075553/
     time-coronavirus-chinas-investment-ai-paying-big-way**

26   Marrow, Alexander and Soldatkin, Vladamir. "Putin takes
     coronavirus precautions as Moscow unveils tracking app."
     Technology News, April 1, 2020. **https://www.reuters.com/
     article/us-health-coronavirus-russia-idUSKBN21J4W8**

27   Harari, Yuval Noah. "The world after coronavirus."
     Financial Times, March 20, 2020. **https://www.ft.com/
     content/19d90308-6858-11ea-a3c9-1fe6fedcca75**

28   Jackson, David. "AP Twitter feed hacked; no attack at White
     House." USA Today, April 23, 2013. **https://www.usatoday.
     com/story/theoval/2013/04/23/obama-carney-associated-
     press-hack-white-house/2106757/**

29   Stuster, J Dana. "Syrian Electronic Army takes credit for
     hacking AP Twitter account." Foreign Policy, April 23, 2013.
     **https://foreignpolicy.com/2013/04/23/syrian-electronic-
     army-takes-credit-for-hacking-ap-twitter-account/**

30   Fisher, Max. "Syrian hackers claim AP hack that tipped stock
     market by $136 billion. Is it terrorism?" The Washington
     Post, April 23, 2013. **https://www.washingtonpost.com/
     news/worldviews/wp/2013/04/23/syrian-hackers-claim-
     ap-hack-that-tipped-stock-market-by-136-billion-is-it-
     terrorism/**

31   Aleem, Zeeshan. "Russia-linked hackers are infiltrating
     the US power grid: report." Vox, September 6, 2017.
     **https://www.vox.com/world/2017/9/6/16262198/
     hackers-us-power-grid-russia**

32 Newman, Lily Hay. "The Worst Cybersecurity Breaches of 2018 So Far." Wired, July 9, 2018. **https://www.wired.com/story/2018-worst-hacks-so-far/**

33 Smith, Rebecca. "Russian Hackers Reach U.S. Utility Control Rooms, Homeland Security Officials Say." The Wall Street Journal, July 23, 2018. **https://www.wsj.com/articles/russian-hackers-reach-u-s-utility-control-rooms-homeland-security-officials-say-1532388110**

34 "Ag and Food Sectors and the Economy." United States Department of Agriculture Economic Research Service. **https://www.ers.usda.gov/data-products/ag-and-food-statistics-charting-the-essentials/ag-and-food-sectors-and-the-economy.aspx**

35 Harari, Yuval Noah. Homos Deus.

36 Benioff, Marc. "We must ensure the Fourth Industrial Revolution is a force for good." World Economic Forum, March 24, 2017. **https://www.weforum.org/agenda/2017/03/we-must-ensure-the-fourth-industrial-revolution-is-a-force-for-good/**

37 Muro, Mark and Maxim, Robert and Whiton, Jacob with contributions from Hathaway, Ian. "Automation and Artificial Intelligence: How machines are affecting people and places." Metropolitan Policy Program at Brookings Institution, January 2019. **https://www.brookings.edu/wp-content/uploads/2019/01/2019.01_BrookingsMetro_Automation-AI_Report_Muro-Maxim-Whiton-FINAL-version.pdf**

38 McCullagh, Kevin. "Stop freaking out about robots," Fast Company, December 5, 2018. Illustration. **https://www.fastcompany.com/90276356/stop-freaking-out-about-robots/**

39 Bhaumik, Debasis. "Size of the Pie!" July 1, 2019. **https://www.linkedin.com/pulse/size-pie-debasis-db-bhaumik-pmp-crisc-mba-msc/**

40 "Gross World Product," Wikipedia. **https://en.wikipedia.org/wiki/Gross_world_product**

41   Roser, Max. "Economic Growth." Our World in Data. **https://ourworldindata.org/economic-growth/**

42   "RACE21™ — The Big Picture of Business Transformation." Teck. **https://www.teck.com/news/connect/issue/volume-27,-2019/table-of-contents/race21-the-big-picture-of-business-transformation**

43   "Ideas at Work: Improving the future through innovation and technology." Teck. **https://www.teck.com/media/Tecks-Approach-to-Innovation-and-Technology.pdf**

44   Griswold, Alison. "How Lyft disguises its losses." Quartz, May 9, 2019. **https://qz.com/1615121/lyft-loses-lots-of-money-but-hides-it-with-non-gaap-accounting/**

45   Anthony, Scott D. and Trotter, Alasdair and Schwartz, Evan I. "The Top 20 Business Transformations of the Last Decade." Harvard Business Review, September 24, 2019. **https://hbr.org/2019/09/the-top-20-business-transformations-of-the-last-decade**

46   Ismail, Salim with Malone, Michael S. and Van Geest, Yuri. Exponential Organizations: Why new organizations are ten times better, faster, and cheaper than yours (and what to do about it). Diversion Books: New York, 2014

47   Hensel, Anna. "How Nike is boosting its direct-to-consumer business with tech acquisitions." Digiday, Mary 10, 2019. **https://digiday.com/retail/nike-boosting-direct-consumer-business-tech-acquisitions/**

48   Westerman, George. Leading Digital

49   Meyersohn, Nathaniel. "Nike's digital reboot is working." CNN Money, July 12, 2018. **https://money.cnn.com/2018/07/12/news/companies/nike-snkrs-app-nikeplus/index.html**

50   Lutz, Zachary. "Inside the Nike+ Accelerator: Fueling the quantified-self movement." Engadget, July 26, 2013. **https://www.engadget.com/2013/07/26/inside-the-nike-accelerator/**

51   "Helping to Make Sport a Daily Habit." Nike News, March 21, 2020. **https://news.nike.com/news/nike-digital-health-activity-resources**

[52] Ries, Eric. The Lean Startup: How Today's Entrepreneurs Use Continuous Innovation to Create Radically Successful Businesses. Crown Business: New York, 2011. 14.

[53] Warren, Katie. "Airbnb just announced it expects to go public in 2020. Meet CEO Brian Chesky, who cofounded the company in 2008 to help pay his San Francisco apartment's rent and is now worth $4.2 billion." Business Insider, September 19, 2019. https://www.businessinsider.com/brian-chesky-airbnb-life-career-net-worth-relationship-philanthropy

[54] Bezos, Jeff. "2016 Letter to Shareholders." Amazon, April 17, 2017. https://blog.aboutamazon.com/company-news/2016-letter-to-shareholders

[55] "Forum on Leadership: A Conversation with Jeff Bezos." [16:00 starts conversation on "customer obsession"]. George W. Bush Presidential Center's Forum on Leadership, April 20, 2018. https://www.youtube.com/watch?v=KPbKeNghRYE

[56] Carroll, Jim. Think Big, Start Small, Scale Fast: Stories from the Stage on Disruption, Transformation and the Accelerating Future. OBLIO Press, 2020. https://jimcarroll.com/2010/05/innovation-think-big-start-small-scale-fast/.

[57] Jeffrey, Jonathan. "What Is An Exponential Organization?" Entrepreneur, October 29, 2019. https://www.entrepreneur.com/article/341439

[58] Souza, André. "Welcome to the era of exponential organizations!" LinkedIn Pulse, July 19, 2016. https://www.linkedin.com/pulse/welcome-era-exponential-organizations-andr%C3%A9-souza/

[59] Ismail, Salim. Exponential Organizations.

[60] "Tesla Model 3 electric car orders accelerate to 276,000." BBC Business, April 3, 2016. https://www.bbc.com/news/business-35953817

[61] "The Week that Electric Vehicles Went Mainstream." Tesla, April 7, 2016. https://www.tesla.com/blog/the-week-electric-vehicles-went-mainstream?utm_campaign=Blog_Model3_040716&utm_source=Twitter&utm_medium=social

[62] Bajarin, Tim. "Why TED Matters." Time, March 24, 2014.
      https://time.com/34784/why-ted-matters/

[63] Shema, Hada. "The Impact of TED Talks." Scientific American.
      February 8, 2014. **https://blogs.scientificamerican.com/
      information-culture/the-impact-of-ted-talks/**

[64] Ismail, Salim. "Massive Transformative Purpose As The
      Only Way to Achieve Exponential Growth." Growth
      Institute, May 10, 2018. **https://www.youtube.com/
      watch?v=zdDkFk_Mlkk**

[65] Isidore, Chris. "Amazon didn't kill Toys 'R' Us. Here's what
      did." CNN Business, March 15, 2018. **https://money.cnn.
      com/2018/03/15/news/companies/toys-r-us-closing-
      blame/index.html**

[66] Morgan, Clancy. "Steve Jobs left Apple to start a new
      computer company. His $12-million failure saved
      Apple." Business Insider, August 21, 2019. **https://www.
      businessinsider.com/steve-jobs-12-million-dollar-failure-
      saved-apple-next-2019-8**

[67] The W. Edwards Deming Institute. Quotes.deming.org/3734

[68] Naudé, Wim. "Artificial Intelligence against COVID-19: An Early
      Review."

[69] "Learn How a 'Phygital' Strategy Can Help Your Business."
      Forbes, August 13, 2018. **https://www.forbes.com/sites/
      sap/2018/08/13/learn-how-a-phygital-strategy-can-help-
      grow-your-business/#309f69f5770a/** Ruqaiyah Jaffery, Aug
      13, 2018,

[70] **https://iabspain.es/estudio/
      estudio-anual-de-ecommerce-2018/**

[71] Machuca, José Maria. "What's Phygital in the Customer
      Experience?" WAM Global Growth Agents, November 15,
      2018. **https://www.wearemarketing.com/blog/whats-
      phygital-in-the-customer-experience.html**

[72] Ferguson, Renee Boucher. "Luminar Insights: A Strategic
      Use of Analytics." MIT Sloan Management Review, February
      12, 2014. **https://sloanreview.mit.edu/case-study/
      luminar-insights-2/**

73 "Novartis Annual Report 2012," Novartis, 14. https://www.
    novartis.com/sites/www.novartis.com/files/novartis-
    annual-report-2012-en.pdf

74 Jobs, Steve. "Stanford University Commencement Address
    2015," June 12, 2005. Transcript published in the

Stanford Report, 14 June 2005. https://news.stanford.edu/
    news/2005/june15/jobs-061505.html

75 "Information Technology: Value Creator or
    Commodity?" Knowledge@Wharton, The Wharton
    School, University of Pennsylvania, March 25, 2004.
    https://knowledge.wharton.upenn.edu/article/
    information-technology-value-creator-or-commodity/

76 Manyika, James and Sneader, Kevin. "AI, automation, and
    the future of work: Ten things to solve for." McKinsey Global
    Institute, June 2018. https://www.mckinsey.com/featured-
    insights/future-of-work/ai-automation-and-the-future-of-
    work-ten-things-to-solve-for#part4

77 "2018 Edelman Trust Barometer." Global Study, Edelman.
    https://www.edelman.com/sites/g/files/aatuss191/
    files/2018-10/2018_Edelman_Trust_Barometer_Global_
    Report_FEB.pdf

78 George, Bill. "The Truth About Authentic Leaders." Harvard
    Business School, July 6, 2016. https://hbswk.hbs.edu/item/
    the-truth-about-authentic-leaders

79 Bremer, Marcella. "Henna Inam: 'We are Wired for
    Authenticity.'" Leadership and Change Magazine, November
    8, 2016. https://www.leadershipandchangemagazine.
    com/henna-inam-we-are-wired-for-authenticity/

80 Grant, Adam. Give and Take: A Revolutionary Approach to
    Success. Penguin, 2013. https://www.google.com/books/
    edition/Give_and_Take/6lFjl3V7ByoC?hl=en&gbpv=0/

81 Fantz, R. L. "Visual experience in infants: Decreased attention
    familiar patterns relative to novel ones." Science, October
    30, 1964, 146 (Whole No. 3644), 668–670. https://doi.
    org/10.1126/science.146.3644.668

[82] Chouinard, Michelle M. and Harris, P. L. and Maratsos, Michael P. "Children's Questions: A Mechanism for Cognitive Development." *Monographs of the Society for Research in Child Development*, Vol. 72, No. 1, (2007), pp. i, v, vii-ix, 1–129. https://www.jstor.org/stable/30163594?seq=1

[83] Cao, Jerry. "10 Worst Design Failures of All Time." Studio, July 17, 2013. https://www.uxpin.com/studio/blog/10-worst-design-failures-of-all-times/

[84] Fruhlinger, Josh. "Equifax data breach FAQ: What happened, who was affected, what was the impact?" CSO, February 12, 2020. https://www.csoonline.com/article/3444488/equifax-data-breach-faq-what-happened-who-was-affected-what-was-the-impact.html/

[85] Bernard, Tara Siegel and Cowley, Stacy. "Equifax Breach Caused by Lone Employee's Error, Former C.E.O. Says." The New York Times, October 3, 2017. https://www.nytimes.com/2017/10/03/business/equifax-congress-data-breach.html

[86] Egan, Matt. "Equifax CEO Richard Smith is out after stunning data breach." CNN Money, September 26, 2017. https://money.cnn.com/2017/09/26/investing/equifax-ceo-richard-smith-out/index.html

Lightning Source UK Ltd.
Milton Keynes UK
UKHW041454030720
365982UK00005B/1404